Study Guide for
Ingraham and Ingraham's

INTRODUCTION TO
MICROBIOLOGY
A CASE HISTORY APPROACH

Third Edition

Jay M. Templin
Montgomery County Community College

D1501741

THOMSON

BROOKS/COLE

Australia · Canada · Mexico · Singapore · Spain
United Kingdom · United States

Printed in the United States of America
1 2 3 4 5 6 7   07 06 05 04 03

Printer: Patterson Printing Company

ISBN: 0-534-39466-3

For more information about our products,
contact us at:
Thomson Learning Academic Resource Center
1-800-423-0563

For permission to use material from this text,
contact us by:
Phone: 1-800-730-2214
Fax: 1-800-731-2215
Web: http://www.thomsonrights.com

Brooks/Cole—Thomson Learning
10 Davis Drive
Belmont, CA 94002-3098
USA

Asia
Thomson Learning
5 Shenton Way #01-01
UIC Building
Singapore 068808

Australia/New Zealand
Thomson Learning
102 Dodds Street
Southbank, Victoria 3006
Australia

Canada
Nelson
1120 Birchmount Road
Toronto, Ontario M1K 5G4
Canada

Europe/Middle East/South Africa
Thomson Learning
High Holborn House
50/51 Bedford Row
London WC1R 4LR
United Kingdom

Latin America
Thomson Learning
Seneca, 53
Colonia Polanco
11560 Mexico D.F.
Mexico

Spain/Portugal
Paraninfo
Calle/Magallanes, 25
28015 Madrid, Spain

CONTENTS

PREFACE

This study guide contains a series of features to facilitate your learning of not only the facts but also the concepts and processes of microbiology. These features are:

Chapter Outline: The outline for the corresponding chapter of the textbook is found at the beginning of each chapter of the study guide. It lists the major topics that you will cover.

Key Terms: A list of the most important words from the text chapter immediately follows the chapter outline.

Study Tips: These are practical tips and suggestions on how to make your study habits and style more effective and efficient.

Study Exercises: Each study guide chapter presents a series of exercises. Each exercise contains questions that relate to a topic in the textbook. The topic is listed at the beginning of the exercise. This reference will help you to locate the relevant information in the textbook as you use the study guide.

Each exercise offers a series of questions that can be answered as you study the corresponding section in the textbook. The questions vary in format, ranging from short answer to matching to true/false. The questions provide an opportunity to actively practice as you study the content in the textbook. They are designed to promote your learning of microbiology. In this edition, four correlation questions have also been included in each chapter to sharpen your analytical and critical thinking skills in studying microbiology concepts.

Multiple Choice: The study exercises are followed by multiple-choice questions that provide a general review of the chapter's content.

Answer Key: The chapter concludes with the correct answers for each of the exercises in the study guide. After answering the questions, you can check this key to learn about your progress toward mastery of the content in the textbook.

Jay M. Templin
Montgomery County Community College
Pottstown, PA

Chapter 1
The Science of Microbiology

Key Terms

genomics

microbiology

pathogen

eukaryote

prokaryote

archaebacteria

algae

bacteria

fungi

protozoa

virus

helminth

spontaneous generation

endospore

immunity

vaccination

attenuated

chemotherapy

selective toxicity

synthetic drug

antibiotic

genetic engineering

recombinant DNA technology

bioremediation

Chapter 1

Study Tips

1. To start studying Chapter 1, scan the chapter for major topics and subtopics. Also, read the chapter summary before studying the chapter in depth. This approach can provide a quick overview of the chapter's content.

2. Research some of the topics from Chapter 1 through the Internet. Key search terms that you can use include: genomics, spontaneous generation, immunity, and other words from the Key Terms list.

3. Learning vocabulary is a major challenge for any student in a science course. After studying Chapter 1 write your own brief definition for each key term in the chapter.

4. Answer the review questions at the end of the chapter. Your instructor can supply you with the correct answers once you have tried to answer the questions on your own.

5. Take some study time to look over all of the chapters in your textbook. Can you find where each topic noted in Chapter 1 is covered in more detail? Where is the topic of recombinant DNA technology covered again? Are there other discussions related to medical microbiology? Can you find a more in-depth description about the differences between eukaryotic and prokaryotic cells?

6. Have you taken other biology courses before microbiology? Can you relate some of the concepts of these courses to the content you are studying now? For example did you learn about the immune system when studying human anatomy and physiology? Where is insulin normally produced in the body? Where does *E. coli* normally inhabit the human body?

7. Expand on this idea of relating microbiology to other courses you have taken or plan to take. How can a course in genetics relate to topics in microbiology? Also, think about the relevance of courses in chemistry and physics. Can the facts and concepts of chemistry help you understand microbial metabolism? Can your findings in physics help you understand topics such as heat and light? How are these topics of physics related to topics in microbiology?

Correlation Questions

1. Read the Case Histories at the beginning of Chapter 1. What symptoms were checked in the patients to assess their physical condition? How does knowledge of microbiology pertain to understanding some of these symptoms?

2. For the Case Histories in this chapter, could other tests have been conducted for a more thorough examination of the patients? Are there other kinds of blood tests in addition to a total white blood cell count? How, for example, could a differential white blood cell count (e.g., percentage of neutrophils) have offered additional information about the patients' immune systems and the kinds of microbes infecting them? Look ahead at Chapters 16 through 18 in the text, dealing with the topic of immunity, to gain some insights for answering this question.

The Unseen World and Ours

A. Label each one of the following statements as true or false. If false, correct the statement.
   1. The bubonic plague was caused by a bacterium.
   2. Infected rats and fleas carried the microorganism that caused the bubonic plague in Europe during the Middle Ages.
   3. Today, cases of the bubonic plague do not occur in the western hemisphere.
   4. The potato blight in Ireland in the 1800s was caused by a bacterium.
   5. In the 1500s, the Native Americans in Mexico had not developed immunity to smallpox and measles when first exposed to the microorganism causing it.
   6. Tetanus and gas gangrene are viral diseases.
   7. A pathogen is a microorganism that has a beneficial effect on humans.

B. Complete each of the following statements with the correct term or terms.
   1. The study of how microorganisms affect the earth and its atmosphere is called _____ microbiology.
   2. Vaccines have been developed to protect humans against _____, microorganisms that can cause disease.
   3. AIDS is an abbreviation for _____.
   4. Using microorganisms to produce a food supply is a goal of _____ microbiology.
   5. The first human use of microbes was to make and preserve _____.
   6. Microorganisms that are used to kill insects are natural _____.
   7. Water is made safe for human consumption through the advances of _____ microbiology.

The Scope of Microbiology

A. Complete each of the following statements with the correct term.
   1. Microorganisms are currently divided into _____ subgroups.
   2. The _____ are the recently recognized subgroup of microorganisms.
   3. A primary distinction among the subgroups of microorganisms is in their _____ structure.
   4. Viruses are particles of the chemical _____, packaged in a protein coat.
   5. Some worm species, the _____, are traditionally a part of the study of microbiology.

B. Match each description to the correct group of organisms.
   1. phytoplankton are members   A. archaea
   2. most are unicellular, prokaryotic with   B. algae
      a uniform, grainy material inside their cells
   3. worms of the animal kingdom   C. bacteria
   4. nucleic acid core, protein coat   D. fungi
   5. one member causes malaria   E. helminths

6. one type causes AIDS
7. most form a mycelium
8. spheres and rods are characteristic shapes
9. most complex unicellular structure
10. one kind causes syphilis
11. kelp belongs to this group
12. discovered in the 1970s

F. protozoans
G. viruses

A Brief History of Microbiology

A. Complete each of the following statements with the correct term or terms.

1. The most powerful magnification of Leeuwenhoek's microscopes was _____X.
2. Leeuwenhoek referred to the microorganisms he studied as _____.
3. Leeuwenhoek communicated detailed drawings of his microorganisms to the _____.
4. The main idea of spontaneous generation was that life can arise from a _____.
5. Redi conducted experiments testing the theory of _____.
6. Pasteur tested the theory of spontaneous generation by using flasks with _____ necks.
7. _____ are structures that some bacteria form, making them resistant to heat.
8. Pasteur's experiments disproving spontaneous generation proved that microorganisms could be studied by rational _____ means.
9. Koch developed four _____ to prove that a particular microorganism causes a disease.
10. _____ _____ is the bacterium causing anthrax.
11. A _____ contains a weakened form of a microorganism that can cause a disease.
12. Jenner is most noted for studying the disease _____.
13. The term "vacca" is a Latin term meaning _____.
14. An attenuated strain of a microorganism is a _____ form of this microbe.
15. Chemotherapy is the treatment of a disease with a _____.

Microbiology Today

A. Label each one of the following statements as true or false. If false, correct it.

1. Pasteur articulated the principle of selective toxicity.
2. Salvarsan was the first drug used to successfully treat syphilis.
3. Erythromycin was the first medically useful antibiotic.
4. Iwanowsky observed that bacteria could pass through the pores in the filters of his experiment.
5. The tobacco mosaic virus infects plants.
6. Bacteria are easy to culture.
7. Bacteria can multiply rapidly.
8. *Staphylococcus aureus* is the microorganism most often used in DNA recombinant studies.

9. Under proper conditions bacteria can double their numbers every 20 minutes.

10. Recombinant DNA experiments led to the technology to produce penicillin.

Multiple Choice: Review

1. Lister soaked bandages in _____. This served as an antiseptic.
   A. alcohol
   B. heat
   C. penicillin
   D. phenol

2. *Yersinia pestis* is the causative agent of the disease
   A. bubonic plague.
   B. pneumonia.
   C. syphilis.
   D. typhoid fever.

3. Which one of the following subgroups has a eukaryotic, complex cell structure?
   A. archaea
   B. bacteria
   C. protozoa
   D. viruses

4. Select the incorrect statement about viruses.
   A. They can live outside a host.
   B. They contain a nucleic acid as part of their structure.
   C. They contain a protein as part of their structure.
   D. They lack a cellular structure.

5. Select the incorrect statement about eukaryotic cell structure.
   A. A true nucleus is lacking in the cells.
   B. Algae have this kind of structure.
   C. Bacteria lack this kind of structure.
   D. Many organelles are present.

6. Select the incorrect characteristic about algae.
   A. eukaryotic
   B. photosynthetic
   C. some are unicellular
   D. some are fun

7. Select the incorrect statement about protozoans.

    A. Cilia are used by some for movement.
    B. Most are unicellular.
    C. Their cells have numerous organelles.
    D. They do not cause any parasitic diseases.

8. Select the incorrect association.

    A. Ehrlich – contributed to the development of chemotherapy
    B. Iwanowski – studied viruses
    C. Koch – developed postulates
    D. Pasteur – proved the theory of spontaneous generation

9. Bacillus anthracis causes

    A. a disease in algae.
    B. a disease in cattle.
    C. smallpox.
    D. syphilis.

10. The first drug named the wonder drug was

    A. erythromycin.
    B. penicillin.
    C. tetracycline.
    D. sulfa drug A.

Answers

Correlation Questions

1. Symptoms examined included body temperature, body areas where pain occurred, pelvic discharge, and results from testing the patients' blood. White blood cells are the infection fighters of the body. A total white blood cell count normally ranges from 5,000 to 10,000 per cubic millimeter. An elevated white cell count, such as 23,000 in one of the patients, indicates an expected response in the human body. More of these cells are needed to fight off the infecting microbe.

2. There are different kinds of white blood cells with different, specific functions. Therefore, more neutrophils might indicate a response to a specific kind of microbe, such as a bacterium. An increase in the percentage of another kind of white blood cell could indicate combating a virus. Other blood cell tests, such as a red blood cell count, can indicate the oxygen-carrying power of the blood. A low red blood cell count can indicate a possible anemia.

The Unseen World and Our World

A.
1. True
2. True
3. False; The bubonic plague still occurs in the western and southwestern United States.
4. False; The potato blight was caused by a fungus.
5. True
6. False; Tetanus and gas gangrene are caused by bacteria.
7. False; A pathogen infects the human body and causes a disease.

B.
1. environmental 2. pathogens 3. acquired immuno-deficiency syndrome 4. agricultural 5. food
6. pesticides 7. environmental

The Scope of Microbiology

A. 1. six 2. archaea 3. cell 4. nucleic acid 5. Helminths

B. 1. B 2. C 3. E 4. G 5. F 6. G 7. D 8. C 9. F 10. C 11. B 12. A

A Brief History of Microbiology

A. 1. 266 2. animalcules 3. Royal Society of London 4. nonliving source 5. spontaneous generation
6. curved 7. endospores 8. scientific 9. postulates 10. *Bacilllus anthracis* 11. vaccine
12. smallpox 13. cow 14. weakened 15. chemical called a drug

Microbiology Today

A.
1. False: Ehrlich articulated the principle of selective toxicity.
2. True
3. False: Penicillin was the first medically useful antibiotic.
4. False: Iwanowsky observed that viruses could pass through the pores in his filters.
5. True
6. True
7. True
8. False: *Escherichia coli* is the microorganism most often used in DNA recombinant studies.
9. True
10. False: Penicillin was discovered before the advent of DNA recombinant technology.

Multiple Choice: Review

1. D 2. A 3. C 4. A 5. A 6. D 7. D 8. D 9. B 10. B

Chapter 2
Basic Chemistry

The Basic Building Blocks
    Atoms
    Elements
    Molecules
Chemical Bonds and Reactions
    Covalent Bonds
    Ionic Bonds
    Hydrogen Bonds
    Chemical Bonds
Water
    Special Properties of Water
    Water Water as a Solvent
Organic Molecules
    Carbon Atoms
    Functional Groups
Macromolecules
    Proteins
    Nucleic Acids
    Polysaccharides
    Lipids
Summary

Key Terms

atom
element
neutron
proton
electron
atomic number
atomic weight
isotope
energy shell
molecule
molecular formula
molecular weight
mole
Avogadro's number
chemical bond
covalent bond
ionic bond
cation

spheres of hydration
colloid
turbid
acid
base
salt
pH scale
buffer
organic molecule
hydrocarbon
functional group
macromolecule
polymer
monomer
dehydration synthesis
hydrolysis
protein
amino acid

anion
hydrogen bond
chemical reaction
reactant
product
free energy
activation energy
catalyst
enzyme
ribozyme
catalytic site
specific heat
solution
solute
solvent
hydrophobic
hydrophilic
dissociate

peptide bond
primary structure
alpha helix
beta sheet
denaturation
sterilization
nucleic acid
nucleotide
purine
pyrimidine
polysaccharide
monosaccharide
disaccharide
lipid
fatty acid
phospholipid
sterol

Study Tips

1. Answer the several kinds of questions at the end of this chapter. Your instructor has the answers to these questions. You can check your mastery of the facts and concepts in this chapter after trying these questions and checking your answers.

2. Good students also learn to write additional questions, anticipating the kinds of questions that may be posed on exams. Trying writing your own questions from each section in the chapter. Try working with another student in your class and testing each other with your questions. One example is to form an answer bank with four groups of biological molecules as choices: carbohydrates, lipids, proteins, and nucleic acids. Next, write a list of descriptions about molecules. Match each description to the type of molecule described. Amino acids are the building blocks of _____. Complete this with the correct choice.

3. Try this as another practice exercise. Attempt to write questions as you anticipate future quizzes and exams in your microbiology course. Write out a list of numbers. One example of a list is: 1, 2, 3, 4, 5, 6, 12, 24, 45, etc. Each number represents a potential answer to a question written from the content of Chapter 2. For example, the numbers 1 through 4 can be valences of atoms. How many covalent bonds can oxygen form? The answer is two. These kinds of questions and answers represent the quantitative concepts stated throughout the chapter.

4. Many colleges supply stickball models and kits to simulate the molecules you are studying in this chapter. Ask your instructor about their availability at your school and use them as another tool to visualize the bonding patterns discussed in Chapter 2.

5. Look through the upcoming chapters in the textbook. Can you find examples where the facts and principles of chemistry will apply to other topics of microbiology that you will study? Examples include metabolism, nutrition, and genetics.

Chapter 2

6. Do you see applications of chemistry to the lab component of your course? What kinds of molecules do bacteria use as a source of energy in their cells? What is meant by the fermentation of carbohydrates?

Correlation Questions

1. The atomic weight of carbon in the periodic table is 12.001. Why is it not listed as 12.000?

2. How does the computation of pOH depend on the changing concentration of a free ion in solution?

3. Is there one element that can have the same atomic number and atomic mass? Explain

4. An element has an atomic number of 19. Name the element. Will it tend to form cations or anions in solution?

The Basic Building Blocks

A. Complete each of the following statements with the correct term or terms.
1. The _____ is the subatomic particle found in energy shells outside the nucleus of the atom.
2. The _____ is the subatomic particle with a positive charge.
3. The _____ is the subatomic particle with no charge.
4. The innermost shell of an atom can hold __(number)__ or __(number)__ electrons.
5. An intact atom has the same number of protons and _____.
6. The atomic _____ is the number of protons and neutrons.
7. Oxygen has an atomic number of _____.
8. Oxygen usually has an atomic weight of _____.
9. The most common isotope of carbon has _____ protons and _____ neutrons.
10. The atomic number of chlorine is _____.
11. An element has 11 protons and 12 neutrons in its nucleus. Its atomic weight is _____.
12. An element has 8 protons and 10 neutrons in its nucleus. Its atomic number is _____.
13. An element has an atomic number of 8. The number of electrons in its outermost energy shell is _____.
14. An element has an atomic number of 7. The number of electrons in its innermost energy shell is _____.
15. The molecular weight of water is _____.

B. Label each of the following statements as true or false. If false, correct it.
1. An intact atom is electrically neutral.
2. Electrons are found in the nucleus of an atom.
3. The electron is not found in the nucleus of an atom.

4. The neutron is found in the nucleus of an atom.
5. The electrons in the innermost energy shell of an atom are valence electrons.
6. If an atom has an atomic number of 6, it has 3 valence electrons.
7. The atomic weight of carbon is 6.
8. A mole of water weighs 24 grams.
9. A mole of hydrogen weighs 5 grams.
10. Sulfur is a trace element.

C. Refer to Figure I to answer the following questions.

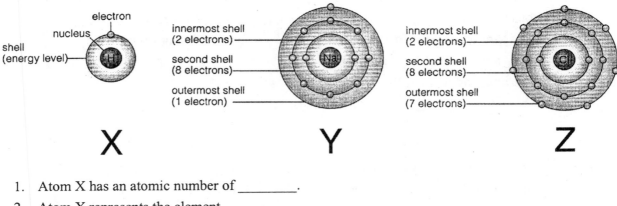

1. Atom X has an atomic number of _____.
2. Atom X represents the element _____.
3. Atom Y has _____ protons in its nucleus.
4. Atom Y represents the element _____.
5. Atom Z has _____ protons in its nucleus.
6. If atom Z has 18 neutrons, its atomic weight is _____.
7. How many more electrons does atom Z need for a full outer shell configuration?
8. Atom Z represents the element _____.

Chemical Bonds and Reactions

A. Label each of the following as describing a covalent bond, ionic bond, or hydrogen bond.
1. It forms between water molecules.
2. It is formed by the activity of an electron donor.
3. It is formed by the sharing of electrons.
4. This bond exists in sodium chloride.
5. It is formed by the activity of an electron acceptor.
6. It can be single, double, or triple.
7. This bond is found in polar and nonpolar molecules.
8. This bond is formed by the attraction between oppositely charged particles.

9. An oxygen atom can form two of this type of bond.
10. This bond stabilizes a large molecule such as DNA, helping to stabilize its shape.
11. This is the most stable bond.
12. This bond involves positive sodium particles.
13. It is an extremely weak bond.
14. NaCl has this kind of bond.
15. Carbon forms four of these bonds.

B. Complete each of the following statements with the correct term or terms.
1. A chemical reaction changes reactants into _____.
2. Reactions only occur between substances with enough energy to enter an _____ state.
3. _____ are substances that increase the rate of a reaction in living systems.
4. _____ are enzyme RNA molecules.
5. Enzyme names end in the suffix _____.
6. *E. coli* has about _____ different enzymes.

Water

A. Label each of the following statements as true or false. If false, correct it.
1. About 40 percent of the weight of a bacterial cell consists of water.
2. Hydrogen bonds increase the heat of vaporization of water.
3. Water is a substance with a high specific heat.
4. As a solid, ice is more dense than liquid water.
5. Compounds that do not interact with water are hydrophilic.
6. Compounds that dissolve in water are hydrophobic.
7. Hydrogen ions have a positive charge.
8. The solution with a hydrogen ion concentration of 0.001 grams per liter has a pH of 4.
9. Acids are proton acceptors in a solution.
10. Buffers are substances that tend to change the pH of a solution.

B. Select the correct choice for each statement.
1. Hydrogen bonds make water
   A. adhesive.
   B. cohesive.
2. NaCl is the _____ of a saltwater solution.
   A. solute
   B. solvent

3. NaCl is

    A. hydrophilic.

    B. hydrophobic.

4. Colloid suspensions are usually

    A. clear.

    B. cloudy.

5. Water-fearing substances are

    A. hydrophilic.

    B. hydrophobic.

6. The charge of a hydroxide ion is

    A. negative.

    B. positive.

7. Acid solutions contain more _____ ions.

    A. hydrogen

    B. hydroxide

8. A basic solution has a pH _____ than 7.

    A. higher

    B. lower

9. The pH of 4 is _____ times more acidic that a pH of 6.

    A. 10

    B. 100

10. The pH of 6 is _____ times more basic than a pH of 3.

    A. 100

    B. 1000

11. Hydrophilic means

    A. water fearing.

    B. water loving.

12. NaCl is

    A. water fearing.

    B. water loving.

13. The more acidic pH is

    A. 5.4.

    B. 5.6.

14. The more basic pH is

    A. 8.4.

    B. 8.8.

15. Some bacteria have an intracellular pH as high as

    A. 7.2.

    B. 8.5.

Organic Molecules

A. Complete each of the following statements with the correct term.
   1. Organic compounds contain at least the elements _____ and _____.
   2. The functional group -COOH is called the _____ group.
   3. The amino group contains the elements _____ and _____.
   4. Molecules that contain the amino groups are called _____, as they readily accept protons.
   5. A polar functional group makes a compound more _____ in water.

Macromolecules

A. Complete the following statements about proteins with the correct term or terms.
   1. The monomers of proteins are _____ _____.
   2. The polymers of proteins are broken into monomers through _____, the addition of water at their chemical bonds.
   3. Amino acids are linked by _____ bonds.
   4. A _____ is two amino acids joined together.
   5. The L and D forms are different _____ of an amino acid.
   6. _____ is the destruction of a protein's three-dimensional structure.
   7. The _____ structure of a protein is its order of bonded amino acids.
   8. Disulfide bonds establish a protein's _____ structure.
   9. The _____ structure of a protein refers to how the different chains of molecules fit together.
   10. The alpha helix and pleated sheet refer to a protein's _____ structure.

B. Refer to Figure II to answer the following questions.

1. Name all of the different elements found in the amino acid.
2. Each peptide bond is formed with the loss of a molecule of _____.
3. The amino acids are joined to form a _____ chain.
4. What does the R group represent on each amino acid?

C. Complete each of the following statements about nucleic acids with the correct term.
   1. Part of a ribosome's composition consists of the nucleic acid _____.
   2. _____ is the nucleic acid that encodes the genetic information in a cell.
   3. _____ are the monomers of nucleic acids.
   4. Cytosine and thymine are _____, single-ringed bases.
   5. Adenine and guanine are _____, double-ringed bases.
   6. _____ is the major carrier molecule of cellular energy.
   7. Nucleotides are joined in a single-stranded nucleic acid by _____ bonds.
   8. DNA has the shape of a _____ helix.
   9. In DNA the base adenine always bonds to _____.
   10. In DNA the base cytosine always bonds to _____.
   11. In DNA the base thymine always bond to _____.
   12. In DNA the base guanine always bonds to _____.

D. Match each of the following descriptions to the correct kind of carbohydrate.
   1. glucose is one example
   2. cellulose is one example
   3. reserve energy storage
   4. reserve energy storage molecule in animals
   5. two monosaccharides bonded
   6. lactose is an example
   7. starch is an example
   8. galactose is an example
   9. simplest sugar
   10. ring form can be alpha or beta
   11. the largest carbohydrate
   12. simple sugar

   A. monosaccharide
   B. disaccharide
   C. polysaccharide

E. Refer to Figure III to answer the following questions.

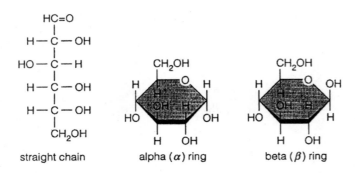

straight chain   alpha (α) ring   beta (β) ring

1. How many carbon atoms are in a molecule of the straight chain?
2. How many hydrogen atoms are in a molecule of the straight chain?
3. How many oxygen atoms are in a molecule of the straight chain?
4. The straight chain represents a molecule of _____.
5. The alpha and beta rings each represent a molecule of _____.
6. If the two ring forms bond, there is a loss of a molecule of _____ between them.

F. Label each of the following statements about lipids as true or false. If false, correct it.
1. The building blocks of lipids are chemically broken down by a dehydration synthesis.
2. Lipids are polar molecules.
3. Fatty acids can be saturated or unsaturated.
4. A phospholipid molecule contains three fatty acids.
5. Fatty acids are joined to glycerol in the lipid molecule by a ester linkage.
6. The phosphate group forms the nonpolar head of a phospholipid.
7. Sterols contain fatty acids.
8. Cholesterol is a sterol.
9. Phospholipid molecules tend to form membranes.
10. Lipids do no make up the genetic systems of cells.
11. Fatty acids with double bonds are unsaturated.
12. Most bacteria have sterols.

Multiple Choice: Review

1. An element has an atomic number of 13 and an atomic weight of 27.The number of protons in its nucleus is
    A. 13.
    B. 14.
    C. 27.
    D. 28.

2. An atom has an atomic number of 13.   The number of electrons in its outermost energy shell is
   A. 2.
   B. 3.
   C. 5.
   D. 7.

3. Select the best description for the water molecule.
   A. sharing of electrons, polar
   B. sharing of electrons, nonpolar
   C. transfer of electrons, polar
   D. transfer of electrons, nonpolar

4. Select the incorrect characteristic about water.
   A. Each of molecule consists of three atoms.
   B. It is a solvent.
   C. It has a boiling point of 212°C.
   D. It is found in the cells of microorganisms.

5. Base molecules dissociate to form _____ ions.
   A. hydrogen
   B. hydroxide

6. A pH of 4 is _____ times more acidic than a pH of 6.
   A. 10
   B. 100
   C. 1000
   D. 10000

7. Select the correct association.
   A. amino acids/polysaccharides
   B. glucose/nucleotide
   C. monosaccharide/glucose
   D. RNA/lipid

8. Select the element found in proteins that is not found in glucose.
   A. carbon
   B. hydrogen
   C. nitrogen
   D. oxygen

9. Select the correct statement about DNA.
   A. It contains the sugar ribose.
   B. It is usually single-stranded.
   C. It makes RNA.
   D. Some of its nucleotide contain uracil.

10. The monosaccharides of sucrose are
    A. glucose and glucose.
    B. glucose and fructose.
    C. lactose and galactose.
    D. lactose and fructose.

Answers
Correlation Questions

1. The value of 12.001 is a weighted average of all isotopes of carbon. Although the most common isotope has an atomic weight of 12, there are rare isotopes of this element with atomic weights of 13 and 14.

2. pOH depends on the concentration of hydroxide ions in a solution. If the concentration of this ion is 0.00000001 (grams per liter), this value is 10 to the negative 8 power. This minus eight exponent converts to a pOH of 8.

3. The most common isotope of hydrogen has one proton and lacks neutrons. Therefore both its atomic number and atomic weight are one.

4. The element is potassium. Its electron arrangement, from the innermost to the outermost energy level, is: 2 -8 - 8 - 1. It tends to lose its single, outermost electron and form a cation.

The Basic Building Blocks

A. 1. electron 2. proton 3. neutron 4. one or two 5. electrons 6. weight 7. eight 8. sixteen 9. six, six 10. seventeen 11. twenty-three 12. eight 13. six 14. two 15. eighteen

B. 1. True
   2. False; Electrons are in the energy shells around the nucleus.
   3. True
   4. True
   5. False; The electrons in the outermost shell of the atom are the valence electrons.
   6. False: It has four valence electrons.
   7. False; The atomic number of carbon is six.
   8. False; A mole of water weighs eighteen grams.
   9. False; A mole of hydrogen weighs two grams.
   10. True

C. 1. one 2. hydrogen 3. eleven 4. sodium 5. seventeen 6. thirty-five 7. one 8. chlorine

Chemical Bonds and Reactions

A. 1. hydrogen 2. ionic 3. covalent 4. ionic 5. ionic 6. covalent 7. covalent 8. ionic 9. covalent 10. hydrogen 11. covalent 12. ionic 13. hydrogen 14. ionic 15. covalent

B. 1. products 2. activation 3. enzymes 4. ribozymes 5. ase 6. one thousand

Water

A. 1. False; Water is about 70 percent of the weight of bacterial cells.
   2. True
   3. True
   4. False; Ice is less dense than liquid water.
   5. False; Compounds that do not interact with water are hydrophobic.
   6. False; Compounds that dissolve in water are hydrophilic.
   7. True
   8. False; This solution has a pH of 3.
   9. False; Acids are proton donors in solution.
   10. False; Buffers tend to stabilize the pH of a solution.

B. 1. B 2. A 3. A 4. B 5. B 6. A 7. A 8. A 9. B 10. B 11. B 12. B 13. A 14. B 15. B

Organic Molecules

A. 1. carbon and hydrogen 2. carboxyl 3. nitrogen and hydrogen 4. bases 5. soluble

Macromolecules

A. 1. amino acids 2. hydrolysis 3. peptide 4. dipeptide 5. isomers 6. denaturation 7. primary 8. tertiary 9. quaternary 10. secondary

B. 1. at least carbon, hydrogen, oxygen, nitrogen 2. water 3. polypeptide 4. The R group is the variable part of the amino acid.

C. 1. RNA 2. DNA 3. nucleotides 4. pyrmidines 5. purines 6. ATP 7. phosphodiester 8. double 9. thymine 10. guanine 11. adenine 12. cytosine

D. 1. A 2. C 3. C 4. C 5. B 6. B 7. C 8. A 9. A 10. A 11. C 12. A

E. 1. six 2. twelve 3. six 4. glucose 5. glucose 6. water

F. 1. False; They are broken down by hydrolysis.
   2. False; Lipids are nonpolar molecules and do not dissolve in water.
   3. True

4. False; A phospholipid molecule contains two fatty acids.

5. True

6. False; The phosphate group forms the polar head of the molecule.

7. False; Sterols lack fatty acids.

8. True

9. True

10. True

11. True

12. False; Most bacteria lack sterols

Multiple Choice: Review

1. A 2. B 3. A 4. C 5. B 6. B 7. C 8. C 9. C 10. B

Chapter 3
The Methods of Microbiology

Viewing Microorganisms
    Properties of Light
    Microscopy
    The Compound Light Microscope
    Wet Mounts
    Stains
    Light Microscopy: Other Ways to Achieve Contrast
    Light Microscopy: Scanning Microscopes
    Other Light Microscopes
    Electron Microscopy
    Scanned-proximity Probe Microscopes: Viewing Atoms and Molecules
    Molecules
    Uses of Microscopy
Culturing Microorganisms
    Obtaining a Pure Culture
    Isolation
    Growing a Culture
    Preserving Cultures
Summary

Key Terms

electromagnetic waves
gamma rays
light
reflection
transmission
absorption
diffraction
refraction
magnification
resolution
numerical aperture
compound light microscope
wet mount
vital stain
heat fixation
simple stain
differential stain

special stain
electron microscope
freeze-fracturing
freeze-etching
scanning tunneling microscopy
atomic force microscopy
pure culture
filtration
incubation
fastidious
sterilization
autoclave
serial dilution
complex medium
differential medium
enrichment culture
stock culture

Study Tips

1. Continue to scan each chapter for major topics and subtopics before studying the chapter content in depth.

2. After studying the chapter, list the boldfaced terms from the textbook and write a definition for each term in your own words.

3. Use the key terms from the text as search terms for additional study of chapter topics through the Internet. Microscopy is a good search term to use first.

4. Outlining the chapter in your own words is another means of active practice to learn the content of the chapter. Start to make a list of topics and subtopics. Under each topic and/or subtopic list several key ideas that you have learned.

5. Answer the various kinds of questions at the end of each textbook chapter. Your instructor can supply you with the correct answers once you have tried to answer the questions on your own.

6. Use your experiences in lab to help you understand the concepts you are learning in lecture. Use of the microscope will help you to understand the function of its parts as well as the concepts of magnification and resolution.

Correlation Questions

1. Read about the topic of osmosis in Chapter 4. If bacterial cells are suspended in a 5 percent NaCl solution, how will this new environment change the volume of their cells? How will this change the ability to view the cells through a phase-contrast microscope?

2. A bacterial population in a test tube of nutrient broth has a concentration of 15,000 cells per ml. Outline the steps of a serial dilution that will lower the cells concentration for easy colony counts if one ml of the final dilution is spread out on a nutrient agar surface in a petri dish.

3. Read some of the main ideas in Chapter 5, dealing with bacterial metabolism. How do you think a knowledge of this topic pertains to the topics on culturing microorganisms?

4. Look over the table of contents in the textbook. Can you find other text topics that will relate specifically to the topics on culturing microorganisms?

5. A given antibiotic can have different effects on Gram negative and Gram positive bacteria. Can you explain this difference? Read the effects of drug action in Chapter 21 on pharmacology.

Viewing Microorganisms

A. Label each of the following statements as true or false. If false, correct it.
   1. Gamma rays are a visible part of the electromagnetic spectrum.
   2. Blue light has longer wavelengths than red light.
   3. Yellow and green are intermediate wavelengths of light.

4. A solution of red dye absorbs the red component of white light.
5. At room temperature pure water has a refractive index of 1.0.
6. Magnification by a lens is increased by making the lens more concave.
7. When light is transmitted it bounces off the surface of an object.
8. Diffraction is the straightening out of light rays.
9. As light enters a denser medium, it slows down.
10. The refractive index of water is greater than the refractive index of air.

B. Short Answer
1. Name the two ways that magnification can be increased in a compound light microscope.
2. How is the contrast of a microscopic image created?
3. _____ is the distinguishing of two points of an image.
4. List the three factors that increase the resolving power of a microscope.
5. How does the wavelength of light affect the resolving power by a microscope?
6. What kind of microscopy utilizes the attractive and repulsive forces between atoms to form an image of them ?
7. What are the lenses used in a compound light microscope?
8. What is the function of the condenser of the compound light microscope?
9. How does changing the objective lens change the total magnification of the compound light microscope?
10. What is the advantage of low power over high power when viewing specimens with the microscope?

C. Matching
1. acid-fast stain
2. Leifson stain
3. Gram stain
4. hanging drop mount
5. negative stain
6. wet mount

A. simplest way to prepare a specimen
B. iodine used as a mordant
C. sulfuric acid used to decolorize
D. used to reveal bacterial capsules
E. used to reveal flagella
F. uses a seal of petroleum jelly

D. Short Answer
1. List the four steps of the Gram stain, in their correct order.
2. List the four steps of the acid-fast stain, in their correct order.
3. List the five steps of the flagellar stain, in their correct order.

E. Match each of the following descriptions to the correct kind of microscopy.
1. operates on principle of light scattering
2. used to reveal antibodies

A. TEM
B. SEM

3. uses a pair of prisms   C. Normarsky

4. uses freeze-fracturing   D. fluorescence

5. can magnify only up to 10,000X   E. darkfield

Cultivating Microorganisms

A. Label each of the following statements as true or false. If false, correct it.

1. A mixed culture contains many different kinds of microorganisms.
2. Sterilization is the elimination of all microorganisms.
3. Moist heat is more effective at killing microorganisms than dry heat.
4. At 15 pounds per square inch water boils at 100 degrees Celsius.
5. Glass and metal instruments can be sterilized by dry heat.
6. Filtration usually removes viruses from liquids.
7. By filtration the filters to remove microbial cells usually have pores of 0.8 micrometers.
8. Sodium chloride is bleach.
9. The serial dilution is used to increase the concentration of microbial cells in a sample.
10. The streak-plate method is used in a test tube of nutrient broth.
11. A suspension with 1200 cells per milliliter is diluted by 3 steps of a serial dilution. Each step dilutes the cell number by one-tenth. After the three steps the cell concentration is 1.2 cells per milliliter.
12. From question #11, two steps from the original suspension produce a cell concentration of 12 cells per milliliter.

B. Match each description to the correct kind of growth medium.

1. used to culture fastidious microorganisms   A. complex
2. isolates endospore-forming bacteria from a large, complex population   B. defined
3. MacConkey agar is one example   C. differential
4. blood agar is one example   D. enrichment
5. favors the growth of some microorganisms and suppresses others   E. selective
6. medium with only a pH indicator is one example   F. selective-differential
7. nutrient agar is one example
8. nutrient broth is one example
9. *E. coli* grows in a simple kind of medium.
10. used to isolate *Salmonella typhi*
11. used for *S. pyrogenes*
12. exact chemical composition known

C. Complete each of the following statements with the correct term or terms.

1. Most bacteria grow over a temperature range of _____ degrees Celsius.
2. *Escherichia coli* lives in the human _____.
3. *E. coli* grows best at a temperature of _____ degrees Celsius.
4. Fungi grow best at a pH range of _____ to _____.
5. Facultative anaerobes can grow in the _____ or _____ of oxygen.
6. Anaerobes cannot grow in the presence of _____.
7. _____ cultures are maintained for study and reference.
8. _____ is freeze-drying.
9. _____ is the removal of all water when frozen.
10. *Treponema pallidum* is the causative agent of _____.
11. _____ *leprae* causes leprosy.
12. Viruses are grown in _____ cultures.

Multiple Choice: Review

1. Select the color with the longest wavelengths from the visible part of the electromagnetic spectrum.
   A. blue
   B. green
   C. red
   D. yellow

2. If all of the light is transferred to an object when it strikes it, the light is
   A. absorbed.
   B. reflected.
   C. translated.
   D. transmitted.

3. The resolving power of a microscope is determined in part of the refractive index of material
   A. between the objective lens and specimen.
   B. between the ocular and objective lens.
   C. in the lenses.
   D. in the specimen.

4. The primary stain for the Gram stain is
   A. acetone.
   B. alcohol.
   C. gentian violet.
   D. safranin.

5. The decolorizing agent for the acid-fast stain is
   A. carbolfuchsin.
   B. heat.
   C. methylene blue.
   D. sulfuric acid.

6. Which kind of microscopy uses two prisms?
   A. darkfield
   B. fluorescence
   C. Nomarsky
   D. phase-contrast

7. The requirements for sterilizing most microorganisms in the autoclave is
   A. 100°C, 15 minutes
   B. 100°C, 20 minutes
   C. 121°C, 15 minutes
   D. 121°C, 20 minutes

8. A sample of broth culture of bacteria is diluted 1000 times. One ml of this diluted culture forms 19 colonies on a nutrient agar surface. The original concentration of cells per ml was
   A. 1900
   B. 19000
   C. 190000
   D. 1900000

9. SPS agar is a _____ medium.
   A. complex
   B. hazardous
   C. differential
   D. selective

10. The purpose of lyophilization is to
    A. add water to an environment.
    B. hold the temperature constant.
    C. remove all moisture.
    D. stabilize the pH.

Answers

Correlation Questions

1. The cells will shrink in the 5 percent NaCl environment outside the cells. This will make the density of the cells increase. This will change the direction and speed that light passes through them. The salty extracellular environment will also make this outside area denser, also changing light passage through it. If the cells shrink significantly, light passage will mainly change through the cells. This will increase contrast with the area outside the cells.

2. Five cells per unit volume is too few. Five hundred is too many. Outline a procedure for a concentration somewhere between these two extremes.

3. The metabolic ability of a microorganism determines the chemical ingredients it can make independently and, therefore, not required in a nutrient medium. A microorganism that can make most chemical ingredients (e.g., vitamins, minerals, etc.) can survive on a minimal medium. The topics of microbial nutrition and genetics also tie in with the culturing of microbes.

4. The differences in the responses of bacteria to the Gram stain depend on differences in the chemical composition of bacterial cell walls. The strategy of some antibiotics is to interfere with the development of a bacterial cell wall. Based on differences in the makeup of the cell wall, some bacteria are more vulnerable to this strategy than others.

Viewing Microorganisms

A. 1. False; Gamma rays have the shortest wavelengths and are invisible.
2. False; Blue light has the shortest, visible wavelengths.
3. True
4. False; It reflects the red wavelengths and this is what you see.
5. False; At room temperature pure water has a refractive index of 1.33.
6. False; Magnification is increased by making the lens more convex.
7. False; When light is transmitted it passes through the surface of an object.
8. False; Diffraction is the bending of light rays.
9. True
10. True

B. 1. Magnification is increased by making the lens more convex or by bringing the object closer to the lens.
2. Some parts of a viewed image absorb light more than others. These differences produce the contrast.
3. resolution
4. Resolution increases by increasing the size of the lenses, increasing the refractive index, and using immersion oil.
5. Smaller wavelengths increase the resolving power.
6. It is called atomic force microscopy.
7. The lenses used in the compound light microscope are the ocular lens and several objective lenses.
8. The condenser concentrates and directs light through the stage mounting the specimen.
9. The ocular lens is constant, usually 10x. The objective lens used can be several choices: 4x, 10x, or 40x. Total magnification is ocular times objective: i.e., 10x times 40x.

10. Use of a low power does not provide maximum magnification but does allow scanning of most or all of the specimen being viewed.

C. 1. C   2. E   3. B   4. F   5. D   6 . A

D. 1. The specimen is stained with a primary stain such as gentian violet. Iodine is applied as a mordant. A decolorizing agent is added. The specimen is counterstained with safranin.
   2. The specimen is stained with the primary stain carbolfuchsin. The microslide is heat-dried. The decolorizing solution of sulfuric acid in ethanol is added. The specimen is counterstained with methylene blue.
   3. A bacterial suspension is fixed with formalin. The slide is air-dried without heating. A mixture of tannic acid and rosaniline dyes is added to the slide. The excess stain is washed off the slide by flooding with water. The slide is air-dried.

E. 1. E   2. D   3. C   4. A   5. B

Cultivating Microorganisms

A. 1. True
   2. True
   3. True
   4. False; At 15 pounds per square inch water boils at 121 degrees Celsius.
   5. True
   6. False; Viruses pass through the pores of the filter.
   7. False; The pores are about 0.45 micrometers.
   8. False; Sodium hypochlorite is bleach.
   9. False; A serial dilution reduces the cell population to a countable number when plated on a nutrient agar surface.
   10. False: The streak plate method is done on a nutrient agar surface in a Petri dish.
   11. True
   12. True

B. 1. B   2. D   3. F   4. C   5. E   6. C   7. A   8. A   9. B   10. E   11. C   12. B

C. 1. 40   2. intestine   3. 374.   4. 5 to 6.0   5. presence or absence   6. oxygen   7. stock
   8. lyophilization   9. sublimation   10. syphilis   11. *Mycobacterium*   12. tissue

Multiple Choice: Review

1. C   2. A   3. A   4. C   5. D   6. C   7. C   8. B   9. D   10. C

Chapter 4
Prokaryotic and Eukaryotic Cells

Structure and Function
The Prokaryotic Cell
    Structure of the Bacterial Cell
    Structures Outside the Envelope
    Cytoplasm
    Endospores
The Eukaryotic Cell
    Structure
Membrane Function
    Simple Function
    Osmosis
    Facilitated Diffusion
    Active Transport
    Group Translocation
    Engulfment
Summary

Key Terms

organelle
nucleoid
envelope
appendage
Gram-negative bacterium
Gram-positive bacterium
mycoplasmas
phagocytosis
lipopolysaccharide
lipid
porin
lipoprotein
binding protein
periplasm
cell wall
peptidoglycan
murein
bacillus
coccus
spirillum
vibrio
transpeptidase
lysozyme

flagellin
monotrichous
amphitrichous
lophotrichous
peritrichous
chemotaxis
aerotaxis
phototaxis
magnetotaxis
axial filament
cytoplasm
ribosome
endospore
sporulation
germination
mitochondrion
chloroplast
cytoskeleton
nucleus
gamete
endoplasmic reticulum
Golgi apparatus
simple diffusion

turgor
cytoplasmic membrane
transmembrane protein
fluid mosaic model
pilus
pilin
fimbria
adhesin
flagellum

osmosis
facilitated diffusion
active transport
group translocation
endocytosis
phagocytosis
pinocytosis
exocytosis

Study Tips

1. Have you discovered some study techniques that work best for you? Does scanning the chapter for major topics and subtopics help? Have you tried outlining the chapter content in your own words? Consider writing definitions for key terms in your own words if you have found mastery of vocabulary a major challenge.

2. Answer the several kinds of questions at the end of this chapter and other chapters to test your mastery of the chapter content.

3. Sketch a typical prokaryotic cell. Label the major parts and describe the function of each.

4. Sketch a typical eukaryotic cell. Label the major parts and describe a function of each.

5. You can organize some of the information from questions #3 and #4 another way. List the cell parts and write their functions. Use this format.

Cell Structure   Function

6. There are many practical applications of diffusion in your everyday life. How many can you list? For example, how do the gas molecules from a perfume bottle spread throughout the atmosphere in a room? How do the molecules in a drop of food coloring change their distribution when placed in a beaker of water?

7. Test your understanding of cell transport by answering questions to the following situation. A eukaryotic animal cell has a solute concentration inside of 1.0% with 99% water. The outside environment is 0.5% solutes and 99.5% water. The cytoplasmic membrane of the cell is permeable mainly to the water and not to the solutes.

In which direction will water diffuse in this situation?

How is osmosis demonstrated in this situation?

How will the animal cell change volume over time?

Would a plant cell or bacterial cell experience the same changes as the animal cell in this situation?

8. There are many practical applications of osmosis in your everyday life. How many can you list? For example, why gargle with saltwater when your vocal cords are swollen from an infection? Why not use pure water?

Correlation Questions

1. A bacterial cell cannot carry out engulfment. A white blood cell in the human body can carry out this process. What adaptations in a white blood cell allow it to do this? How does this ability of a white blood cell protect the human body?

2. Consider this lab situation. Your lab instructor informs you that he will present you with a series of preserved slides in the lab for labeling. In each case you will be asked to label the cell shown as eukaryotic or prokaryotic. How will you study and prepare to answer these lab questions?

3. Consider this situation. You have an upper respiratory tract infection. A sample of sputum is taken from this area of your body, in order to identify the infecting microorganism. How will this knowledge help with recommending the correct prescription to treat your infection?

4. As a cell becomes larger and larger, its surface to volume ratio changes. How does this change limit the size of a cell?

The Prokaryotic Cell

A. Label each one of the following statements as true or false, If false, correct it.
1. The term "prokaryotic" means true nucleus.
2. Prokaryotic cells are usually larger than eukaryotic cells.
3. Gram-negative bacteria have a complete cell envelope with all three layers.
4. Under ideal conditions a cell of E. coli can divide every 20 minutes.
5. Gram-positive bacteria are usually more resistant to antibiotics than Gram-negative bacteria.
6. In Gram-negative bacteria the cell wall lies in the periplasm.
7. Teichoic acids add to the integrity of Gram-positive cell walls.
8. Bacilli always divide in only one plane.
9. Turgor pressure develops in a bacterial cell as it loses water.
10. Mycoplasmas have a complex cell wall.
11. The hydrophilic tail of a phospholipid molecule consists of two fatty acid chains.
12. The function of the bacterial cell pilus is mainly locomotion.
13. A bacterium avoids an area with high oxygen by aerotaxis.
14. Eukaryotic ribosomes are 70S.
15. Endospores have a high water content.

B. Match each of the following descriptions to the correct cell structure.

1. capsule
2. cell wall
3. endospore
4. flagellum
5. nucleoid
6. outer membrane
7. pilus
8. ribosome

A. can have a peritrichous arrangement
B. composed of murein
C. appendage for cell attachment
D. has proteins called porins
E. DNA area without a defined membrane
F. 70S in bacteria
G. outer, slimy layer of cell envelope
H. structure resistant to heat

C. Complete each of the following statements with the correct term or terms.

1. The small size of cells permits a _____ growth rate.
2. Smaller cells have a higher surface-to-_____ ratio than larger cells.
3. Gram-_____ bacteria lack the outer membrane of the cell envelope.
4. The main function of the cell capsule is _____.
5. The lipopolysaccharide of the outer membrane has a hydrophilic _____ and a hydrophobic _____.
6. _____ are bacteria that lack a cell wall.
7. _____ and _____ are the two sugars with alternating units in the peptidoglycan of the bacterial cell wall.
8. The _____ is the sphere-shaped bacterial cell.
9. Autolysins are _____ that break the cross-linking bonds in peptidoglycan molecules.
10. The _____ of a bacterial cell withstands turgor pressure, preventing the bursting of the cell.
11. Mycoplasmas do not burst by pumping _____ out of their cells.
12. Normal bacteria can be converted to L forms by treating the normal cells with the enzyme _____.
13. The fluid mosaic model refers to the _____ of the bacterial cell.
14. Another word for pilus is _____.
15. A species of the genus _____ causes gonorrhea.
16. Peritrichous refers to an arrangement of _____ around a bacterial cell.
17. By _____ bacteria swim to a region of maximum light intensity.
18. The cytoplasm of the bacterial cell is about 90 percent _____.
19. Antibiotics do not harm the _____ ribosomes of eukaryotic cells.
20. Endospores can germinate to new _____.

The Eukaryotic Cell

A. Label each one of the following statements as true or false. If false, correct it.
1. Bacterial cells are eukaryotic.
2. Eukaryotic flagella produce locomotion by a rapid, rotating motion.
3. The cells of fungi lack a cell wall.
4. The cell walls of plants consist of cellulose.
5. Microtubules are the largest threadlike proteins that compose the cytoskeleton.
6. The nuclear envelope is defined by a single membrane.
7. Eukaryotic DNA is chemically different from prokaryotic DNA.
8. Mitosis is the kind of cell division that produces gametes.
9. The smooth ER is covered with ribosomes.
10. Lysosomes package materials in eukaryotic cells.

B. Match each of the following structures to its correct description.
1. cell wall   A. composed of threadlike proteins
2. chloroplast   B. contains thylakoids
3. cilium   C. repackages new vesicles
4. cytoskeleton   D. synthesizes phospholipids
5. Golgi apparatus   E. makes proteins
6. lysosome   F. resists turgor pressure
7. RER   G. has enzymes for digesting substances
8. SER   H. cell appendages for locomotion

C. Label each one of the following statements as describing a prokaryotic or eukaryotic cell.
1. The cell contains many kinds of organelles.
2. Human cells are this type.
3. The cell diameter is one to three micrometers.
4. Nuclear envelope is absent.
5. Ribosomes are 70S.
6. Ribosomes are 80S.
7. Cell undergoes meiosis.
8. They can be Gram positive or Gram negative.
9. The bacterial cell is this type.
10. Mitochondria and the ER are present.

Passage of Materials Across Cell Membranes

A. Complete each one of the following statements with the correct term or terms.

1. _____ molecules are not soluble in water and pass through the cytoplasmic membrane by dissolving through it.

2. Water passes through the cytoplasmic membrane by passing through _____ in the membrane.

3. By a concentration _____ more molecules are found in one region compared to another.

4. The cell does not expend _____ for the simple diffusion of molecules across cytoplasmic membranes.

5. _____ is the diffusion of water across a membrane.

6. From plasmolysis a cell changes volume by _____.

7. By _____ diffusion molecules are transported from a region of higher concentration to a region of lower concentration. Carrier molecules mediate this process.

8. By _____ _____ molecules are located from a region of lower concentration to a region of higher concentration. The cell spends energy for this process.

9. _____ _____ is a variation of active transport conducted only by bacterial cells.

10. By _____ solid material is engulfed by a cell.

11. By _____ liquid material is taken in by the cell.

12. _____ is the expulsion of material from the cell.

Multiple Choice: Review

1. Select the correct statement about prokaryotic cells.
   A. Human cells are prokaryotic cells.
   B. Most are smaller than eukaryotic cells.
   C. Their nucleus has a defined membrane.
   D. They contain many kinds of organelles.

2. Syphilis is caused by a species of
   A. *Bacillus*.
   B. *Escherichia*.
   C. *Pseudomonas*
   D. *Treponema*.

3. The _____ is the rod shape of bacterial cells.
   A. bacillus
   B. coccus
   C. spirillum
   D. vibrio

4. A bacteria cell that has a higher salt concentration inside compared to its outside environment will

    A. attract water and develop a turgor pressure.
    B. attract water and lose a turgor pressure.
    C. lose water and develop a turgor pressure.
    D. lose water and lose a turgor pressure.

5. By aerotaxis bacteria swim toward a

    A. favorable source of oxygen.
    B. magentic field.
    C. source of glucose.
    D. source of light.

6. Which cell appendage is used for attachment by bacterial cells?

    A. cilium
    B. flagellum
    C. pilus
    D. pseudopodium

7. The 70S-80S ribosome difference accounts for the

    A. effectiveness of some antibiotics.
    B. development of a turgor pressure.
    C. differences in the rate of cell division.
    D. storage ability of cells.

8. Which cell organelle conducts protein synthesis?

    A. lysosome
    B. Golgi complex
    C. mitochondrion
    D. RER

9. Each of the following describes simple diffusion except

    A. develops from the random movement of molecules
    B. energy must be spend by cells for it to occur
    C. molecules spread out
    D. tends to produce an equilibrium

10. Osmosis is the movement of

    A. salts through a membrane.
    B. salts without a membrane.
    C. water through a membrane.
    D. water without a membrane.

Answers
Correlation Questions

1. The rigid cell wall of a bacterium prevents engulfment. The absence of a cell wall, along with a flexible cytoplasmic membrane, allows a white blood cell to surround and enclose things through phagocytosis. If it engulfs a pathogen invading the human body, the white blood cell can destroy the invader and protect the human body.

2. A prokaryotic cell is smaller and simpler compared to a eukaryotic cell. The prokayotic cell also has several distinctive shapes.

3. Is your infection viral? In that case an antibiotic will not work. A bacterial infection can be treated by an antibiotic.

4. Consider a cube-shaped cell. If one of its edges equals one, its surface area is six (6 times 1). Its volume is 1 x 1 x 1. At this size its surface to volume ratio is 6 to 1. Compute its area if each side of the cube is two. Its surface is 24 (6 x 4). Its volume is 8 (2 x 2 x 2). The surface to volume ratio has decreased to three (24 divided by 8). How does this ratio change if each side of the cell is 3? Can you explain how the ratio of surface to volume changes as the cell becomes bigger? How does this limit cell size?. Remember that the cytoplasmic membrane of a cell defines its area. What is the function of this membrane?

The Prokaryotic Cell

A. 1. False: The term "prokaryotic" means before a nucleus. The term "eukaryotic" means true nucleus.
2. False; A eukaryotic cell is usually about ten times larger than a prokaryotic cell.
3. True
4. True
5. False; Gram-negative bacteria are usually more resistant to antibiotics compared to Gram-positive bacteria.
6. True
7. True
8. True
9. False: Turgor pressure builds up in a bacterial cell as it gains water.
10. False; Mycoplasmas are the only group of bacteria that lack a cell wall.
11. True
12. False; The function of the pilus for the bacterial cell is usually attachment.
13. False; By aerotaxis a bacterium swims toward a source of oxygen.
14. False; Eukaryotic ribosomes are 80s. Prokaryotic ribosomes are 70s.
15. False; The water content of the bacterial endospore is less than fifteen percent.

B. 1. G 2. B 3. H 4. A 5. E 6. D 7. C 8. F

C. 1. rapid 2. volume 3. positive 4. protection 5. hydrophilic head, hydrophobic tail
6. mycoplasmas 7. NAG, NAM 8. coccus 9. enzymes 10. cell wall 11. sodium ions
12. lysozyme 13. cytoplasmic membrane 14. fimbria 15. Neisseria 16. flagella
17. phototaxis 18. water 19. 80S 20. vegetative cells

The Eukaryotic Cell

A. 1. False; Cells of the bacteria and archaea are prokaryotic.
2. False; Eukaryotic flagella produce motion by a slow, undulating motion.
3. False: Fungal cells have a cell wall.
4. True
5. True
6. False; The nuclear envelope is defined by a double membrane.
7. False; Eukaryotic and prokaryotic DNA are identical.
8. False; Mitosis produces the body cells. Meiosis produces gametes.
9. False; The rough ER is covered with ribosomes.
10. False; The Golgi apparatus packages materials in the eukaryotic cell. The lysosome releases enzymes to break down substances.

B. 1. F 2. B 3. H 4. A 5. C 6. G 7. E 8. D

C. 1. eukaryotic 2. eukaryotic 3. prokaryotic 4. prokaryotic 5. prokaryotic 6. eukaryotic
7. eukaryotic 8. prokaryotic 9. prokaryotic 10. eukaryotic

Passage of Materials Across Cell Membranes

A. 1. hydrophobic 2. pores 3. gradient 4. energy 5. osmosis 6. collapsing 7. facilitated 8. active transport 9. group translocation 10. phagocytosis 11. pinocytosis 12. exocytosis

Multiple Choice: Review

1. B 2. D 3. A 4. A 5. A 6. C 7. A 8. D 9. B 10. C

Chapter 5
Metabolism of Microorganisms

Key Terms

metabolism

catabolism

precursor metabolite

ATP

reducing power

biosynthesis

polymerization

assembly

oxidation

reduction

dehydrogenation reaction

hydrogenation

free energy

substrate level phosphorylation

chemiosmosis

electron transport chain

electron acceptor

aerobic respiration

glycolysis

TCA cycle

pentose phosphate pathway

anaerobic metabolism

fermentation

autotroph

heterotroph

chemotroph

phototroph

cyclic photophosphorylation

oxygenic photosynthesis

anoxygenic photosynthesis

allosteric inhibition

end-product inhibition

Study Tips

1. Answer the several kinds of questions at the end of this chapter, and other chapters, to test your mastery of key facts and concepts.

2. Outline the broad patterns of glycolysis, aerobic respiration, and photosynthesis from the details of the chapter. Focus on the formation of ATP molecules within your outline.

3. Can you relate the lessons of this chapter to your current experiences in lab? For example, if a bacterium ferments glucose in a Durham tube, what produces the color change in the growth medium from red to yellow? What is the source of gas accumulation in the inverted tube of this setup?

4. Bacteria are excellent models to study patterns of metabolism that apply to other kinds of organisms. Can you find some examples of this in the chapter? Begin with lactic fermentation in animal tissues. What is the source of fatigue in skeletal muscles? What is meant by the oxygen debt that must be repaid after intense exercise by the human body?

5. Continue to use the boldfaced terms in the text as search terms for additional study on the Internet.

Correlation Questions

1. Consider growing each kind of microbe in two different environments. In one environment, oxygen is abundant. In the other, oxygen is completely absent. All other characteristics of the two environments are identical, such as the exact composition of the growth medium in the petri dishes. In each case, how will the growth rate of the two microbes compare in the two environmental settings?

   A. strict anaerobe
   B. strict aerobe
   C. facultative anaerobe
   D. aerotolerant microbe

2. A bacterium is studied for its ability to synthesize a substance that is part of coenzyme structure. Microorganism A cannot make it. Microorganism B can make it. Why is the substance considered a vitamin in A and not in B?

3. A microorganism stops making an amino acid after several hours of growth, due to endproduct inhibition. How does this means of metabolism regulation differ from allosteric activation? How are the two processes similar?

4. How is the genetics of microorganisms related to their metabolism? Read ahead in Chapter 6, dealing with bacterial genetics, to help you understand the connection between metabolism and genetics.

Metabolism: An Overview

A. Complete each of the following statements with the correct term.
   1. _____ refers to all biochemical reactions that take place in a cell.
   2. If *E. coli* grows aerobically, this means in the presence of _____.
   3. Only twelve organic compounds, _____ metabolites, are needed as starting points for the metabolic reactions in *E. coli*.

4. Through _____ the cell makes small molecules from precursor metabolites.

5. Polymers such as DNA and proteins are the large molecules or_____ of the cells.

6. By the process of_____ macromolecules are used to make cell organelles.

Aerobic Metabolism

A. Label each of the following statements as true or false. If false, correct it.

1. Porins are small, water-filled holes in the outer membrane of *E. coli*.

2. Most molecules pass through porins by active transport.

3. Transporters are enzymes that bring nutrients into the cell.

4. Substances enter the bacterial cell by translocation through simple diffusion.

5. Normally *E. coli* transports fructose-6-phosphate into the cell and converts it to glucose-6-phosphate.

6. Reduction is the loss of electrons.

7. During oxidation a proton is usually transported along with an electron.

8. The cell uses its reducing power to make ATP.

9. The reduced form of NAD(P) is NAD(P)+.

10. More stable reactant bonds and more instable product bonds drive a chemical reaction.

11. Most phosphate-containing metabolic intermediates contain high-energy bonds to make ATP.

12. During chemiosmosis the cell uses a proton gradient to make ATP from ADP and phosphate.

13. Cytochromes of the electron transport chain in *E coli* accept electrons and hydrogen ions.

14. Eukaryotic organisms generate six ATP molecules for each pair of electrons passing through the electron transport chain.

15. Central metabolism begins with the sugar sucrose.

16. By glycolysis two molecules of glucose are produced from one molecule of pyruvate.

17. During glycolysis, there is a net gain of six ATP molecules per glucose molecule without the contribution of substrate level phosphorylation.

18. Acetyl CoA enters the TCA by combining with a six-carbon molecule, oxaloacetate.

19. Production of NADH and NADPH increases the reducing power in a cell.

20. *E. coli* produces about 10,000 different enzymes used in its biosynthesis.

21. *E. coli* can make all 20 amino acids required to build proteins.

22. The ordering of amino acids in a protein is determined by the information in the structure of DNA.

23. To make glycogen in *E. coli* , the glucose building block is first converted to glucose-6-phosphate.

24. Glucose is a polymer made from glycogen.

25. Peptidoglycan is the major macromolecule of the outer membrane of the bacterial cell.

Anaerobic Metabolism

A. Complete each of the following statements with the correct term.
1. _____ anaerobes are capable of both aerobic and anaerobic metabolism.
2. In anaerobic respiration a substance other than _____ is the final electron acceptor.
3. Fermentation generates fewer molecules of ATP per molecule of substrate than do _____ or _____ respiration.
4. Lactic acid bacteria form __(number)__ molecules of ATP per glucose molecule fermented.
5. In alcoholic fermentation pyruvate is converted to alcohol and _____ .

Nutritional Classes of Microorganisms

A. Match each term to its correct description.

1. autotroph	A. generates ATP and reducing power from chemical reactions	
2. chemotroph	B. light-feeder	
3. heterotroph	C. self-feeder	
4. phototroph	D. different-feeder	

B. Complete each of the following statements with the correct term.
1. Autotrophs make precursor metabolites from the gas _____ .
2. Autotrophs use the pathways of the _____ cycle to make precursor metabolites.
3. Autotrophs combine carbon dioxide with a five-carbon sugar to produce the substance _____ .

C. Label each of the following statements as describing cyclic photophosphorylation or anoxygenic photosynthesis.
1. It does not use two photosystems.
2. It forms NADPH.
3. It generates reducing power.
4. It uses photosystems II and I.
5. It cannot generate reducing power.

Regulation of Metabolism

A. Define each of the following.
1. allosteric enzyme
2. effector
3. end-product inhibition
4. allosteric activation

Multiple Choice: Review

1. Amino acids are bonded into proteins in a cell. This is an example of
   A. assembly.
   B. biosynthesis.
   C. catabolism.
   D. polymerization.

2. Porins in the outer membrane of the bacterial cell are
   A. enzymes that combine with substrate molecules.
   B. substrate molecules requiring molecules.
   C. water molecules that serve as substrates.
   D. water-filled holes in the outer membrane.

3. There are _____ starter compounds, precursor metabolites, in *E. coli*.
   A. six
   B. twelve
   C. twenty-five
   D. fifty

4. Dehydrogenation reactions involve the loss of
   A. electrons.
   B. protons.
   C. electrons and protons.
   D. neither electrons nor protons.

5. Acetyl CoA enters the TCA after being formed from
   A. glucose.
   B. glucose-6-phosphate.
   C. oxaloacetate.
   D. pyruvate.

6. The concentration of _____ is an indicator of a cell's reducing power.
   A. ADP
   B. glucose
   C. NAD(P)H
   D. oxaloacetate

7. In the electron transport chain the quinones accept
   A. electrons.
   B. hydrogens.
   C. electrons and hydrogens.
   D. oxygen.

8. Which organism is the self-feeder?

    A. autotroph
    B. chemotroph
    C. heterotroph
    D. phototroph

9. Effectors control the function of allosteric enzymes by changing their

    A. concentration.
    B. location in the cell.
    C. pH requirement.
    D. shape.

10. By end-product inhibition, the buildup of an end-product _____ the rate of the first reaction of the associated metabolic pathway.

    A. decreases
    B. increases

Answers

Correlation Questions

1. A. More growth occurs in the environment where oxygen is abundant. Oxygen is toxic to a strict anaerobe.

    B. More growth occurs in the environment where oxygen is present. The strict aerobe needs oxygen for growth.

    C. The facultative anaerobe can grow in either setting, but should grow more where oxygen is present and it can make more ATP to run its metabolism.

    D. For the aerotolerant organism, the growth should be the same in the two settings. This organism cannot use oxygen, but it is not harmed by it.

2. A vitamin is a dietary requirement. In either organism the substance is necessary for its metabolism. However, it is a vitamin for A because it cannot synthesize it. It must be furnished premade in its growth medium.

3. By end-product inhibition, there is a buildup of the last substance made in the metabolic pathway. As it accumulates, this end product feeds back to the first enzyme in the pathway, inhibiting its action. Therefore it shuts down the pathway and prevents additional, unneeded synthesis of the product. The end product is an allosteric effector. In allosteric activation, the increase in the concentration of an allosteric effector activates an enzyme in a metabolic pathway, stimulating more production of the end product in the pathway.

4. Read Chapter 6. DNA makes RNA which makes protein. All enzymes consist of protein. The genetic blueprint in a specific DNA molecule dictates the chemical makeup of a specific protein. As enzymes are mainly protein, DNA also dictates the makeup of enzyme structure through this

process. Enzymes control each step of metabolism. Therefore, DNA determines the number and kinds of enzymes in a cell, determining the metabolic abilities of the cell.

Metabolism: An Overview

A. 1. metabolism   2. air or oxygen   3. precursor   4. biosynthesis   5. macromolecules   6. assembly

Aerobic Metabolism

A. 1. True

2. False; Most molecules pass through porins by simple diffusion.

3. True

4. False; Translocation is an energy-requiring process.

5. False; *E. coli* transports glucose into the cell and converts it to glucose-6-phosphate.

6. False; Oxidation is the loss of electrons. Reduction is the gain of electrons.

7. True

8. True

9. False; NAD(P)+ is the oxidized form of NAD(P).

10. False; More instable reactant bonds and more stable product bonds drive a chemical reaction.

11. False; Only a few metabolic intermediates produced by substrate level phosphorylation have the high energy bonds to make ATP.

12. True

13. False; Cytochromes of the electron transport chain accept only electrons.

14. False; Eukaryotic organisms generate three ATP molecules for each pair of electrons passing through the electron transport chain.

15. False; Central metabolism begins with the sugar glucose.

16. False: By glycolysis one molecule of glucose is converted to two molecules of pyruvate.

17. False: During glycolysis there is a net gain of two ATP molecules per glycose molecule without the contribution of substrate level phosphorylation.

18. False; Acetyl CoA enters the TCA by coming with a four-carbon molecule, oxaloacetate.

19. True

20. False; E. coli produces about 2,604 different enzymes. About 1/5 are used in its biosynthesis.

21. True

22. True

23. False; In E. coli, glucose is converted to glucose-1-phosphate to make glycogen.

24. False; Glycogen is a polymer made from the building block glucose.

25. False; Peptidoglycan is the major macromolecule of the cell wall of the bacterial cell.

Anaerobic Metabolism

A. 1. facultative   2. oxygen   3. anaerobic and aerobic   4. two   5. carbon dioxide

Nutritional Classes of Microorganisms

A. 1. C   2. A   3. D   4. B

B. 1. carbon dioxide   2. Calvin-Benson   3. ribulosebisphophate

C. 1. cyclic   2. anoxygenic   3. anoxygenic   4. anoxygenic   5. cyclic

Regulation of Metabolism

A. 1.   An allosteric enzyme is an enzyme that changes activity when bonded to a signal molecule.
2.   The effector is the signal molecule that binds to an allosteric enzyme to change its activity.
3.   The end product of a metabolic pathway binds to the first enzyme in the pathway, inhibiting its activity.
4.   An increase in the intracellular concentration of an effector activates an allosteric protein

Multiple Choice: Review

1. D   2. D   3. B   4. C   5. D   6. C   7. B   8. A   9. D   10. A

Chapter 6
The Genetics of Microorganisms

Structure and Function of Genetic Material
    The Structure of DNA
    Reactions of DNA
    Replication of DNA
    Gene Expression
Regulation of Gene Expression
    Regulation of Transcription
    Regulation of Translation
    Global Regulation
    Two-component Regulatory Systems
Changes in a Cell's Genetic Information
    The Genome
    Mutations
    Selecting and Identifying Mutants
    Use of Mutant Strains
    Genetic Exchange Among Bacteria
Genetic Exchange Among Eukaryotic Microorganisms
Population Dynamics
Summary

Key Terms

DNA	codon
nucleotide	anticodon
replication	mutation
gene expression	genome
transcription	plasmid
translation	genotype
nucleoside triphosphate	phenotype
replication fork	mutagen
semiconservative replication	transposon
replication apparatus	direct selection
antiparallel	indirect selection
DNA polymerase	brute strength
ligase	transformation
mRNA	conjugation
tRNA	transduction

Study Tips

1. Continue to answer the several kinds of questions at the end of each chapter. After answering them, ask your instructor to post the answers.

2. Remember that good students also write their own questions, particularly as they anticipate possible exam questions. Work with another student in class and test each other with your questions.

3. Some colleges provide take-apart models of the DNA double helix. Ask your instructor if they are available at your school. Using them can help you to understand the three-dimensional makeup of DNA. It can also help you to understand how it replicates.

    Through this model make an order of 12 base pairs. You can select the sequence of bases for the molecule you make. Select one strand of the DNA from this model. What is the sequence of RNA bases transcribed by this half of the DNA molecule? Write the sequence of the four amino acids translated by this mRNA strand.

4. One of the important ideas in this chapter is the flow of genetic information in the cell. For additional practice, try the following. Write out a series of DNA base pairs. Here is one example:

    TCC - TAT - GCA - CCG - AGT

    Answer these questions:

    If this is one-half of the DNA double helix, what is the base sequence of the other half of the DNA molecule? Read the stated bases from left to right.

    From the originally-stated sequence of DNA bases, what is the series of bases transcribed into mRNA?

    From these five mRNA codons, what is the series of amino acids transcribed by this sequence?

    What is the anticodon for each tRNA molecule carrying each amino acid to the ribosome?

5. As a variation of question #4, try the following by working through the process backwards. Write five tRNA anticodons from left to right.

    What is the order of the mRNA codons that will attract these five anticodons at the ribosome?

    What amino acid will be placed by each mRNA codon at the ribosome? Write them from left to right.

    What is the sequence of bases of DNA that will transcribe the five mRNA codons? Write them from left to right.

Correlation Questions

1. An initial study of the structure of DNA indicates that it does not have the necessary complexity to encode genetic information. How do you think some observers reach this conclusion?

2. How does DNA have the necessary complexity to function as the genetic material if it consists of only four different building blocks (nucleotides)?

3. The chromosomal DNA in a bacterium has 3 million base pairs. How can this information be used to estimate the number of genes in its genome?

4. In the evolution of the cell, genes control the metabolism of proteins. How is this a successful strategy for controlling metabolism, considering that other kinds of macromolecules exist in addition to proteins?

Structure and Function of Genetic Material

A. Label each one of the following statements as true or false. If false, correct it.
   1. The base in a nucleotide of DNA can be either A, C, G, or T.
   2. The base pairs of DNA are either A-G or C-T.
   3. By translation RNA makes DNA.
   4. The nucleotide of DNA contains two phosphate groups.
   5. During DNA replication, the replication forks move in opposite directions.
   6. The terminus is where the bubble forms that initiates replication of the circular, bacterial chromosome.
   7. The function of DNA ligase is to cleave the two strands of the DNA double helix.
   8. The two strands of the DNA double helix are parallel.
   9. The three kinds of RNA are messenger, transfer, and ribosomal.
   10. During transcription ribonucleoside triphosphates pair with exposed bases on DNA.
   11. The promoter is a structural gene on DNA that transcribes mRNA to make a protein.
   12. The anticodon is a base triplet on mRNA.
   13. Activation occurs when tRNA is linked to an amino acid.
   14. A protein produced by translation corresponds to a single gene.
   15. AUG is a termination codon for transcription.

B. Complete each of the following statements with the correct term or terms.
   1. _____ is the science that studies the heredity of organisms.
   2. The nucleotides of DNA consist of three parts: deoxyribose, phosphate, and a _____.
   3. The base pairs of DNA are united by _____ bonding.
   4. DNA is synthesized from the polymerization of _____, its building blocks.
   5. Deoxynucleotides react with ATP to form nucleotide _____.

6. Each strand of the DNA double helix has a three prime end and a _____ prime end.

7. The replication of DNA is not conservative; it is _____.

8. The type of RNA that carries the amino acid to the ribosome is _____.

9. In prokaryotes transcription is guided by the enzyme _____.

10. Transcription begins at a promoter and ends at a site, the _____.

11. The _____ is the base triplet of tRNA that is complementary to the codon on the mRNA.

12. A _____ codon does not encode for any amino acid.

13. The number of amino acids directed by the genetic code is _____.

14. The _____ - _____ sequence is the ribosome binding site.

15. An mRNA with attached ribosomes and proteins is a _____.

Regulation of Gene Expression

A. Label each of the following as describing inducible enzymes, repressible enzymes, or constitutive enzymes.

  1. They are always produced because they are always needed.
  2. Many are needed to metabolize different sugars.
  3. They are produced only when there product is needed.
  4. The sugar serves as a signal molecule.

B. Label each one of the following statements about the Lac operon as true or false. If false, correct it.

  1. Galactoside permease splits a disaccharide into two monosaccharides.
  2. Beta-galactosidase brings lactose into the cell.
  3. The gene lacI lies within the lac operon.
  4. The lac repressor binds to the lac operator in the absence of lactose.
  5. The effector, allolactose, has a negative effect on the lac repressor.

C. Complete the following statements about attenuation.

  1. Attenuation usually regulates the production of enzymes in _____ pathways for amino acids.
  2. The _____ operon is an attenuation regulated operon.
  3. A/an _____ loop prevents the formation of an attenuator loop.
  4. When more histidine is needed, the cell makes more of the _____ needed to make the amino acid.
  5. Histidine regulates the activity of one of the enzymes to make it by end-product _____.

D. Complete the following statements on global regulation and the two-component regulatory system.

  1. By catabolite _____, many genes and operons are regulated and coordinated in response to a source of carbon for growth.
  2. CAP is a _____ protein.

3. Cyclic _____ is the signal molecule for global regulation.

4. Most studies of global regulation involve the bacterium _____.

5. A sensor is a protein that detects an environmental signal and transmits it to another protein, the _____ _____.

6. The regulation of _____ in *E. coli* illustrates the principles of two-component regulation.

Changes in Cell's Genetic Information

A. Match each of the following terms to its correct description.

1.	beta galactosidase	A.	agent causing genetic change
2.	conjugation	B.	change in cell's genetic makeup
3.	genome	C.	effect of genes on a cell function
4.	genotype	D.	enzyme
5.	mutagen	E.	cell's genetic plan
6.	mutation	F.	process that transfers R factor
7.	phenotype	G.	R factor is one kind
8.	plasmid	H.	sum total of DNA in a cell

B. Label each of the following as describing a chemical mutagen, physical mutagen, or biological mutagen.

1. Hydroxylamine reacts with cytosine.

2. It involves a transposable element.

3. It involves jumping genes.

4. It produces a thymine dimer.

5. It is produced by 5-BU.

6. It is produced by UV light

C. Define each of the following.

1. missense mutation

2. nonsense mutation

3. lethal mutation

4. conditionally expressed mutation

D. Explain the difference between direct selection and indirect selection for identifying mutants.

E. Label each of the following statements as describing transformation, conjugation, or transduction.

1. A virus transfers genes between cells.

2. Hfr cells carry this out.

3. It can be generalized or specialized.

4. It can be natural or artificial.

5. It involves the F plasmid.

6. It is used to map genes on chromosomes.

7. It requires a pilus.

8. Phages with life cycles are involved.

9. Viruses reproduce at the expense of the host cell in the process.

10. DNA is absorbed from the environment.

11. *Streptococcus pneumoniae* has surface enzymes called nucleases that cut bound DNA into fragments for incorporation in cells.

12. A bacteriophage is involved.

Multiple Choice: Review

1. In the DNA double helix the number of thymine bases always equals the number of _____ bases.
   A. adenine
   B. cytosine
   C. guanine
   D. uracil

2. During transcription of DNA a sequence of ATCG will order an RNA base sequence of
   A. ATCG.
   B. GCTA.
   C. GCUA.
   D. UAGC.

3. The function of tRNA is to
   A. build the ribosome.
   B. carrying amino acids to the ribosome.
   C. catalyze nucleotides.
   D. transport nucleotides to DNA.

4. Select the incorrect description about the anticodon.
   A. It is a base triplet.
   B. It is found on mRNA.
   C. It is smaller than mRNA.
   D. It lacks a nucleotide sequence of bases.

5. The role of the repressor in the lac operon is to
   A. bind to the lac operator.
   B. remove lactose from the cell.
   C. stimulate transcription.
   D. transport lactose into the cell.

6. The genome is the
    A. signal for enzyme induction.
    B. signal for enzyme repression.
    C. sum total of all DNA in the cell.
    D. sum total of all RNA in the cell.

7. A nonsense mutation stops translation before a protein product is complete.
    A. True
    B. False

8. UV light is an example of a source of a _____ mutation.
    A. biological
    B. chemical
    C. physical
    D. spontaneous

9. By conjugation
    A. cells reproduce asexually.
    B. DNA replication is stopped.
    C. plasmids are destroyed.
    D. plasmids are transferred to cells.

10. Transduction is carried out by a
    A. alga.
    B. fungus.
    C. protozoan.
    D. virus.

Answers

Correlation Questions

1. An alphabet with four letters can form few words. An incorrect assumption, however, is that the four DNA nucleotides are read (transcribed) one at a time.

2. However, read in groups of three, the number of combinations of DNA bases increases. Four to the third power equals 64.

3. If a researcher knows the average length of a gene (number of DNA base pairs), the number of genes can be estimated by simple division. If an average gene is 1000 base pairs, with 3 million base pairs, there are 3000 genes in this bacterium. This assumes that all of the DNA is transcribed.

4. Proteins have many functions ranging from transport to enzyme activity. Every step of metabolism depends on a protein, as a large part of an enzyme is protein and enzymes run metabolism. Other kinds of macromolecules do not rival proteins for their diversity.

Structure and Function of Genetic Material

A. 1. True
   2. False; The base pairs of DNA are A-T and G-C.
   3. False; By translation RNA makes protein.
   4. False; Each nucleotide of DNA contains one phosphate.
   5. True
   6. False; The terminus is where two completed chromosomes separate after being copied from the original bacterial chromosome.
   7. False; The function of DNA ligase is to seal a segment of DNA into a DNA molecule.
   8. False; The two strands of the DNA double helix are antiparallel.
   9. True
   10. True
   11. False; The promoter signals the binding of RNA polymerase for transcription.
   12. False; The anticodon is a base triplet on tRNA.
   13. True
   14. True
   15. False; AUG is a codon that initiates transcription

B. 1. genetics 2. nitrogen base 3. hydrogen 4. nucleotides 5. triphosphates 6. five prime
   7. semiconservative 8. transfer RNA 9. RNA polymerase 10. terminator 11. anticodon
   12. nonsense 13. twenty 14. Shine-Dalgarno 15. polysome

Regulation of Gene Expression

A. 1. constitutive 2. inducible 3. repressible 4. inducible

B. 1. False; Galactoside permease brings lactose into the cell.
   2. False; Beta-galactosidase splits a disaccharide into two monosaccharides.
   3. False; The gene lacI lies outside the lac operon.
   4. True
   5. True

C. 1. biosynthesis 2. histidine 3. antiterminator 4. enzymes 5. inhibition

D. 1. repression 2. regulatory 3. AMP 4. *E. coli* 5. response regulator 6. chemotaxis

Changes in the Cell's Genetic Structure

A. 1. D 2. F 3. H 4. E 5. A 6. B 7. C 8. G

B. 1. chemical 2. biological 3. biological 4. physical 5. chemical 6. physical

C. 1. It is a change in a codon to one that codes for a different amino acid.

   2. It is a change in a codon to one that stops translation.

   3. It is a change in DNA that stops DNA replication

   4. It is a genetic change rendering a gene product nonfunctional only in certain environments.

D. By direct selection conditions are created that promote the growth of the desired mutant strain. By indirect selection the growth of a desired mutation strain is prevented. A condition is then imposed on growing cells.

E. 1. transduction 2. conjugation 3. tranduction 4. transformation 5. conjugation 6. conjugation
    7. conjugation 8. transduction 9. transduction 10. transformation 11. transformation
    12. transduction

Multiple Choice: Review

1. A   2. D   3. B   4. B   5. A   6. C   7. A   8. C   9. D   10. D

Chapter 7
Recombinant DNA Technology and Genomics

Key Terms

recombination	blunt-end ligation
homologous	transfection
recombinant DNA	microinjection
intron	electroporation
cell extract	probe
reverse transcriptase	shotgun cloning
origin of replication	gene bank
ligation	polymerase chain reaction
restriction endonuclease	gene therapy
rotational symmetry	genomics
palindrome	sequencing
cohesive ends	assembling
anneal	annotation
RFLP	microarray technology
DNA ligase	

Study Tips

1. Continue to answer the several kinds of questions at the end of each chapter. After answering them, ask your instructor to post the answers. Did you make any mistakes? Talk to your instructor and class members in order to help you improve your understanding of the facts and concepts in the chapter.

2. This chapter reveals the potential to treat many human diseases through the administration of products developed by recombinant DNA technology. From your study can you list any human diseases that can be treated through this technology? Start with the following list of products that are being used. Choose one that you find interesting and learn more about it from library research and the Web. What is the function of this product in the body? What human disease develops in its absence?

erythropoietin
clotting factors
growth hormone
insulin
interferon
surfactant
tumor necrosis factor
tPA
vaccines for hepatitis

3. Every day in the national news there are stories that draw from the principles explained in this chapter. Consider the following. A small amount of DNA, equivalent to one gene, is collected at a crime scene. For analysis through DNA fingerprinting, about 75,000 copies of the gene are necessary. By the PCR, how many cycles of replication from this original gene are necessary to make the analysis?

4. Read the newspapers and national magazines, plus science journals, to find other examples of recombinant DNA technology.

Correlation Questions

1. A pair of homologous chromosomes is studied in a eukaryotic cell. One chromosome has a gene sequence of PQRstuv. On the other chromosome of the pair the sequence of corresponding genes is pqrSTUV. How can crossing-over establishing two homologs, one with only dominant genes (capital letters) and one with only recessive genes (lower-case letters)?

2. Why can crossing-over not occur in prokaryotic cells?

3. Without crossing over, how can genetic recombinants occur in bacterial cells?

4. Select one hormone that can be produced by recombinant DNA technology to treat a human disease. How will it change the structure/function of the human body to improve its health?

Recombinant DNA Technology and Genomics

A. Complete each of the following statements with the correct term or terms.
   1. _____ is the process of forming a new combination of genes, natural or artificial.
   2. In eukaryotes, crossing over occurs between _____ chromosomes.
   3. Gene _____ is the process of obtaining a large number of copies of a gene from a single gene.
   4. A cloning _____ is a DNA molecule that a host cell will replicate in the process of gene cloning.
   5. About __(number)__ percent of the dry weight of a bacterial cell consists of DNA.
   6. _____ are noncoding regions on the chromosome of a eukaryotic cell.
   7. Through the enzyme _____ _____, mRNA makes intron-free DNA.

8. By the process of _____ a fragment of DNA is sealed into a chromosome.

9. Restriction endonucleases evolved in bacteria probably to protect them from attack by _____.

10. Taq I is a restriction endonulcease obtained from the organism _____ _____.

11. AGATCT is a DNA base sequence known as a _____.

12. Due to _____ symmetry, DNA bases read the same way in one strand as they do in the opposite direction of the other strand in the DNA double helix.

13. _____ ends are sticky ends of short, complementary, single-stranded regions of DNA produced when type II restriction endonucleases cut DNA asymetrically within the axis of rotational symmetry.

14. When the sticky ends of DNA pair anneal, this means they _____.

15. _____ is a process whereby DNA from a virus is used as a cloning vector.

16. By _____ DNA is introduced into cells when electrical impulses destabilize the plasma membrane.

17. A _____ is a short DNA molecule that is complementary to its corresponding mRNA.

18. A _____ _____ is a set of bacterial clones , each which carries a small clone of DNA.

19. Cells of the organism _____ _____ were the host first used for recombinant DNA.

20. Human cells tend to glycosylate protein molecules after synthesis, meaning that they add _____ molecules to the proteins.

B. Define each of the following.
   1. recombinant molecule -
   2. cloning vector -
   3. cell extract -
   4. reverse transcriptase -
   5. ligation -
   6. transfection -
   7. electroporation -
   8. microinjection -

C. Label each of the following statements as true or false. If false, correct it.
   1. There are 20 to 40 copies of plasmids (pBR type) per cell of *E. coli*.
   2. Recombinant DNA procedures require large amounts of DNA..
   3. PCR can produce a pure solution of a multiplied gene.
   4. To perform PCR, four different kinds of deoxyribonucleoside triphosphates are required.
   5. The enzyme of *Thermus aquaticus*, taq polymerase, is sensitive at very high temperatures.
   6. PCR is sensitive enough to detect one gene from a single virus in a sample of blood.

Genomics

Complete each of the following statements with the correct term or terms.

1.   Genomics is the study of an organism as revealed by the _____ of bases in its DNA.
2.   _____ sequencing does not require extensive special equipment but is slow and tedious.
3.   Only about __(number)__ bases can be sequenced in a single run.
4.   _____ is the computer-based process of searching for presumptive genes.
5.   _____ technology is a means of determining which of an organism's genes are being expressed under a given set of conditions.

Multiple Choice: Review

1.   Select the correct description about cloning.
   A.   The copies of a gene produced in a clone have wide genetic variation..
   B.   It cannot reproduce many copies of a gene.
   C.   It is used in recombinant DNA technology.
   D.   It uses human cells as vectors for bacterial cells.

2.   Reverse transcriptase is an enzyme that directs
   A.   RNA making DNA.
   B.   RNA making protein.
   C.   DNA making DNA.
   D.   RNA making protein.

3.   Introns are
   A.   coding DNA regions in the eukaryotic cell.
   B.   coding DNA regions in the prokaryotic cell.
   C.   noncoding DNA regions in the eukaryotic cell.
   D.   noncoding DNA regions in the prokaryotic cell.

4.   Select the palindromic sequence.
   A.   ATGCGT
   B.   ATAGCG
   C.   CGGCGG
   D.   GCAGAC

5.   Which technique is used to seal together the ends of DNA?
   A.   cloning
   B.   gel electrophoresis
   C.   ligation
   D.   restriction

6. Which process inserts DNA into cells using a virus vector?

   A. electroporation
   B. microinjection
   C. transfection
   D. transformation

7. The effect of hGH in an organism is to

   A. decrease growth.
   B. decrease water absorption.
   C. increase growth.
   D. increase water absorption.

8. Three kilobases contain _____ base pairs.

   A. 3000
   B. 30,000
   C. 300,000
   D. 3,000,000

9. Glycosylation is the

   A. adding of sugars to proteins.
   B. digestion of proteins into amino acids.
   C. fermentation of sugars.
   D. synthesis of DNA from RNA.

10. By hand sequencing, 20,000 to 50,000 DNA bases can be sequenced in about

    A. one year.
    B. 30 seconds.
    C. 10 minutes.
    D. 2 months.

Answers

Correlation Questions

1. If the chromosome break occurs between genes Rs on one homolog and rS on the other homolog, the two new recombinants will be PQRSTUV and pqrstuv. Crossing over is the exchange of corresponding regions of homologous chromosomes.

2. Bacteria do not have homologous chromosome pairs. They are not diploid. They are haploid with one circular chromosome.

3. Genetic recombination can occur with bacteria by conjugation, transduction, and transformation. By bacterial conjugation, for example, there is a recombination of genes to produce a new chromosome.

4. Examples include treating anemia with EPO, a hormone that the human body produces naturally to signal the bone marrow to produce red blood cells. Also, insulin, produced by recombinant DNA technology, can treat diabetes mellitus (type I).

Recombinant DNA Technology and Genomics

A. 1. recombination 2. homologous 3. cloning 4. vector 5. three 6. introns 7. reverse transcriptase
   8. ligation 9. viruses 10. *Thermus aquaticus* 11. palindrome 12. rotational 13. cohesive
   14. join together 15. transfection 16. electroporation 17. probe 18. gene bank 19. *E. coli*
   20. sugar

B. 1. It is a molecule derived from part of one chromosome and part of another.
   2. It is a DNA molecule that a cell will replicate in the cloning process.
   3. It is the liquid content of ruptured cells.
   4. It is an enzyme that uses mRNA to make a complementary strand of DNA.
   5. It is sealing a DNA fragment with a gene, to be cloned, into a cut region of a cloning vector.
   6. It is putting recombinant DNA into a host cell when a virus is the cloning vector.
   7. Cells are suspended in a DNA solution and exposed to high-voltage impulses. This destabilizes the cytoplasmic membranes, making it possible for DNA in solution to enter the cells..
   8. DNA is inserted directly into some animal cells by a pipette when in a vacuum.

C. 1. True
   2. False; They require amounts in only nanograms or micrograms.
   3. True
   4. True
   5. False; This enzyme functions best at high temperatures and can withstand repeated cycles of heating.
   6. True

Genomics

1. sequence 2. hand 3. 500 4. Annotation 5. Microarray

Multiple Choice: Review

1. C 2. A 3. C 4. B 5. C 6. C 7. C 8. A 9. A 10. A

Chapter 8
The Growth of Microorganisms

Key Terms

binary fission	energy source
budding	hydrogen peroxide
doubling time	superoxide
growth rate	cofactor
exponential growth	coenzyme
synchronous culture	thermophile
lag phase	mesophile
log phase	psychrophile
stationary phase	barophile
death phase	alkaliphile
inoculum	plasmolysis
batch culture	halophile
limiting nutrient	centrifugation
chemostat	turbidity
colony	dessicator
biofilm	spectrophotometer
growth yield	Coulter counter
major element	sampling error
trace element	

Study Tips

1. Try the review, correlation, and essay questions at the end of the chapter. After answering them, discuss the results with other students in your class. Are your answers similar?

2. Good students are skillful at writing their own questions, particularly as they anticipate possible exam questions. Some of the possible questions in this chapter involve calculations. Here are some examples to get you started.

    Beginning with one cell, can you estimate the size of a bacterial population after three hours? Assume synchronous growth and a doubling time of twenty minutes.

    A bacterial population in nutrient broth has a concentration of 40,000 cells per ml. Can you plot out a series of tenfold serial dilutions that reduces this concentration to 40 cells per ml for plating on a nutrient agar surface?

3. Do you see applications of the information in this chapter to your lab experiences? What does turbidity in a test tube of nutrient broth indicate? Why, through a serial dilution, is choosing plates between 30 to 300 colonies a good compromise between speed and accuracy? How can several colonies in a mixed culture be separated into pure cultures?

4. This chapter has many applications to the upcoming lessons on the physical and chemical control of microorganisms. Look at Chapter 9 for examples.

Correlation Questions

1. A bacterial population, growing in a test tube of nutrient broth, passes through various phases: lag – log – stationary – death. How are these patterns applicable to the growth states of other kinds of populations, including humans currently populating planet earth?

2. How do the principles of chemistry and physics support the understanding of microbial growth patterns?

3. How do the principles of cell biology support the understanding of microbial growth patterns?

4. Phosphate accumulation in a stream can change the life forms in many ways in that waterway. How, for example, could this change affect the microbial life plus the other species in the stream?

Populations/The Way Microorganisms Grow

A. Label each of the following statements as true or false. If false, correct it.
    1. Microbial growth refers to a change in size of an individual cell.
    2. Most microbial cells divide through mitosis.
    3. A population with a long doubling time grows rapidly.
    4. Cells growing in a bacterial culture are usually not growing synchronously.
    5. The death phase can be a highly dynamic state for a bacterial population..
    6. The inoculum is the cells used to start a culture.
    7. Cells growing on the surface of a colony exposed to air are fully anaerobic.
    8. The vast majority of microbial cells in natural liquid environments are found in assemblages called biofilms.

B. Complete each of the following statements with the correct term.

1. By the process of _____ a bubble-like growth enlarges and separates from a parent cell.

2. Another name for generation time is _____ time.

3. *E. coli* synthesizes about 30 _____ molecules as it enters the stationary phase.

4. A _____ culture is grown in a closed container.

5. By the formula of 2 raised to the n power (exponent), after seven generations one original cell will have produced __(number)__ cells. Assume synchronous growth.

6. Cells on the edge of a plated colony are usually in the _____ phase of growth.

C. Match each of the following descriptions to the correct growth phase of a microbial population. Each description matches to only one letter. Throughout the match a letter can be used more than once.

1. exponential growth occurs   A. death

2. starting, no-growth period   B. lag

3. cells start to die   C. log

4. growth is very slow   D. stationary

5. occurs before exponential growth

6. most rapid growth rate

D. Describe each of the following.

1. binary fission

2. poor medium

3. limiting nutrient

4. chemostat

5. continuous culture

6. colony

What Microorganisms Need to Grow

A. Label each of the following statements as true or false. If false, correct it.

1. *E. coli* grows ten times faster on a poor medium than on a rich medium..

2. Nitrogen is a trace element for bacterial growth.

3. Nitrogenases enable nitrogen-fixing bacteria to use atmospheric nitrogen..

4. All aerobes among bacteria produce superoxide dimutase..

5. Heterocysts are cells that are permeable to oxygen.

6. Hydrogen peroxide is removed from compounds by perioxidases.

7. Phosphorus occurs exclusively in cells as the phosphate ions.

8. Sulfur is about 5 percent of the dry weight of cells.

9. *Leuconostoc citrovorum* is a lactic acid bacterium.

10. Psychrophiles grow best at very high temperatures.

11. Thermophiles grow best at very low temperatures.

12. *E. coli* is a mesophile.

13. Prokaryotes living in the deep ocean are barophiles.

14. Acidophiles thirve at high pH values.

15. If the concentration of solutes increases outside a bacterial cell, if faces the threat of losing water by osmosis.

B. Complete each of the following statements with the term greater or less.

1. The growth rate of a bacterial population is _____ in the exponential phase compared to the lag phase.

2. The rate of cell death is _____ than the rate of cell increase during the lag phase.

3. A rich medium has a _____ concentration of nutrients compared to a poor medium.

4. The concentration of sulfur in a cell is _____ than the concentration of nitrogen.

5. The temperature range for bacterial cell growth is _____ than the range for eukaryotic cells.

6. Thermophiles have a growth rate that is _____ at 0 degrees Celsius compared to the growth rate of psychrophiles at this temperature.

7. The growth of mesophiles is _____ at 37 degrees Celsius compared to 10 degrees Celsius.

8. Proteins have _____ heat sensitivity compared to other kinds of molecules.

9. If the pH is above 7.6 for *E. coli* , the growth of this bacterium is _____.

10. Halophiles have _____ sensitivity to high salt concentrations outside their cells compared to other kinds of cells.

Measuring Microbial Growth

A. Describe how each of the following is used to measure the number of microorganisms.

1. turbidity
2. dry weight
3. metabolic activity

Multiple Choice: Review

1. In a rich laboratory medium, *E. coli* has a doubling time of about _____ minutes.

    A. 2
    B. 4
    C. 18
    D. 54

2. Exponential growth of a population occurs in the _____ phase.

    A. death
    B. lag
    C. log
    D. stationary

3. In a chemostat
   A. the four stages of population growth do not occur.
   B. nutrients are constantly supplied.
   C. the population is subjected to temperature variation. .
   D. the number of microbial cells is always changing.

4. *E. coli* is a(n)
   A. facultative anaerobe.
   B. microaerophile.
   C. obligate aerobe.
   D. obligate anaerobe.

5. Oxygenases
   A. add oxygen directly to organic molecules.
   B. break down an organic carbon source.
   C. build an organic carbon source.
   D. remove oxygen from organic molecules.

6. Phosphorus makes up about _____ percent of the dry weight of microorganisms.
   A. three
   B. seven
   C. twelve
   D. fifteen

7. A bacterium thrives at 90 degrees Celsius. It is a(n)
   A. anaerophile.
   B. mesophile.
   C. psychrophile.
   D. thermophile.

8. A barophile is a _____ lover.
   A. acid
   B. base
   C. pressure
   D. salt

9. An acidophile grows best at a pH of
   A. 4
   B. 7
   C. 9
   D. 11

10. The spectrophotometer detects the _____ of a culture.
    A. chemical content
    B. location
    C. pH
    D. turbidity

Answers

Correlation Questions

1. All populations are subject to the same growth patterns. Fish in a pond or humans on the earth can potentially pass through the stages: lag - log - stationary - death. As bacterial population size is limited by space and nutrients in its confined environment, other kinds of populations are subject to the same factors in their environments.

2. Examples include the concentration and structure of molecules, pH and temperature.

3. Examples include cell division, cell structures, cell metabolism, and osmosis.

4. If phosphate does not remain a limiting factor, algal blooms can occur. This can deplete the waterway of oxygen, affecting microbes plus populations of larger organisms. For example, some species of fish (e.g., trout) require highly- oxygenated water.

Populations/The Way Microorganisms Grow

A. 1. False; Microbial growth refers to a change in the size of the population.
    2. False; Most microbial cells divide through binary fission.
    3. False; A population with a short doubling time grows rapidly.
    4. True
    5. True
    6. True
    7. False; Cells growing on the surface of a colony are fully exposed to air and are aerobic.
    8. True

B. 1. budding 2. doubling 3. protein 4. batch 5. one hundred and twenty-eight 6. exponential

C. 1. C 2. D 3. A 4. B 5. B 6. C

D. 1. This is how a bacterial cell elongates and divides.
    2. This is a growth medium that has a minimal supply of nutrients..
    3. This is the nutrient in scarcest supply for a population, thus setting the growth of a population.
    4. This is a device with an input supplying fresh media to a microbial population. There is an output removing cells and toxins from the population.

5. This is a culture with a constant level of cells due to an input of fresh media and an output for aging cells and toxins.

6. A colony is a solid mass of cells on an agar surface, all produced from the same cell.

What Microorganisms Need to Grow

A. 1. False; Autotrophic microorganisms obtain their carbon from carbon dioxide.

2. False; Chemoheterotrophs obtain their carbon from glucose.

3. True

4. True

5. False; Heterocysts are impermeable to oxygen.

6. True

7. True

8. False; Sulfur is about one percent of the dry weight of microorganisms.

9. True

10. False; Psychrophiles grow best at very low temperatures.

11. False; Thermophiles grow best at very high temperatures.

12. True

13. True

14. False; Acidophiles thrive at low, acidic pH values.

15. True

B. 1. greater 2. less 3. greater 4. less 5. greater 6. less 7. greater 8. greater 9. less 10. less

Measuring Microbial Growth

A. 1. As cell numbers in a population increase, the growth medium becomes cloudier or more turbid. A spectrophotometer can measure this turbidity and estimate cell numbers.

2. The dry weight of a culture can be collected by centrifuging that culture. The weight is divided by the weight per cell, yielding the number of cells.

3. The rate of forming metabolic products, for example, is directly related to cell mass.

Multiple Choice: Review

1. C 2. C 3. B 4. A 5. A 6. A 7. D 8. C 9. A 10. D

Chapter 9
Controlling Microorganisms

Key Terms

sterilization

disinfection

decontamination

antisepsis

microbiocidal

microbiostatic

decimal reduction time

thermal death point

thermal death time

heat

radiation

photosensitizer

photoreactivation

filtration

drying

osmotic strength

chemotherapeutic agent

germicide

germistat

phenol coeffcient

paper disc method

use-dilution test

phenol

alcohol

halogen

tincture

hydrogen peroxide

emulsion

alkylating agent

formalin

temperature

canning

pasteurization

Study Tips

1. Continue to answer the questions at the end of the chapter. Ask your instructor to post the answers once you have tried the questions. Discuss the results with your classmates.

2. Vocabulary is a major part of this chapter and the other chapters in the text. Try this format to organize the vocabulary.

   Term   *Definition*

   List any term that you find difficult to describe. Study the text and then write a definition in your own words.

3. There are several important mathematical concepts in this chapter. Ask your instructor for some practice problems on computations involving the D-value and serial dilution.

4. Can you relate some of the concepts of this chapter to the lab component of the course? Why are most cultures grown for study at 48 hours? How does the autoclave sterilize equipment in your lab?

5. Explore sources on the Internet to learn more about the topics of this chapter. Key search terms include: disinfection, microbiocidal, microbiostatic, etc.

Correlation Questions

1. A population of bacteria in a test tube of nutrient broth consists of 100,000 cells. It is treated with a bacteriocidal agent. This setting is sterile in two hours. What is its D-value?

2. Calculus is an important course taken by the undergraduate major. Why is this background important for you to understand some of the principles of this chapter on the control of microorganisms?

3. Hundreds of years ago Americans preserved meat by salting it and drying it in the sun. How can this be an effective control method?

4. Why does the term "death" have a different meaning for microbes compared to its meaning among multicellular organisms?

The Way Microorganisms Die

A. Complete each of the following statements with the correct term.
   1. _____ is a treatment to destroy all microbial life.
   2. _____ kills microorganisms on living tissue.
   3. If a treatment is microbiostatic it _____ rather than _____ microorganisms.
   4. The D-value is the time in minutes required to kill _____ percent of the cells in a population.

5. During the lethal treatment of microbes, _____-phase cells are more susceptible to microbial death than cells in the stationary phase.

6. _____ _____ is the bacterium causing botulism that is resistant to heat by forming endospores.

7. The TDP is the _____ temperature required to kill all microorganisms in liquid suspension in 10 minutes.

8. The TDT is the _____ time required to kill all microorganisms in a particular liquid at a given temperature.

Physical Controls on Microorganisms

A. Label each of the following statements as true or false. If false, correct it.

1. Ultraviolet light penetrates through an object to kill all microorganisms.

2. The autoclave employs moist heat.

3. The vegetative cells of thermophilic bacteria can withstand prolonged boiling.

4. Low temperatures are sufficient to kill most microorganisms.

5. Long wavelengths of light are lethal to most microorganisms.

6. UV wavelengths are not visible to humans.

7. X rays add electrons to atoms.

8. Ionizing radiation is rarely used for the physical control of microorganisms in the microbiology laboratory.

9. Viruses can normally be removed from liquids by filters in the microbiology lab.

10. Sublimation is the conversion from the gaseous state to the liquid state.

Chemical Controls on Microorganisms

A. Complete each of the following statements with the correct term.

1. A germicide is a chemical that _____ microorganisms.

2. A _____ is a chemical that inhibits microbial growth.

3. At least __(number)__ germicides are sold in the United States..

4. The phenol _____ is the ratio of endpoints between phenol and another germicide.

5. The end point of a germicide uses a dilution of 1:100000 compared to 1:100 for phenol. The germicide is __(number)__ as powerful as phenol.

6. The size of the zone of _____ around a disc on an agar surface is used in the paper disc method.

7. The first noticeable effect of phenol on microbial cells is its disruption of the cytoplasmic cell _____.

8. Phenolics have a _____ group attached to a benzene ring.

9. In a tincture iodine is combined with _____.

10. A __(numbers)__ percent of hydrogen peroxide is a weak antiseptic.

B. Select the one incorrect statement among the choices for each of the following chemical control methods.

1. phenols/phenolics

   A. These substances are denaturing agents.
   B. They are found in certain throat sprays.
   C. Hexachlorophene is currently restricted by prescription.
   D. Hexachlorophene is still used frequently in hospitals.

2. alcohols

   A. These compounds have a hydroxyl group.
   B. Ethanol is an example.
   C. They kill endospores.
   D. They dissolve lipids on the skin..

3 halogens/hydrogen peroxide

   A. Iodine is an example.
   B. They inactivate proteins.
   C. Free chlorine kills microorganisms..
   D. Hydrogen peroxide is formed by catalase.

4. alkylating agents

   A. Ethylene oxide is an alkylating agent.
   B. They remove short carbon chains from proteins.
   C. Formalin is used to embalm tissues.
   D. Low formaldehyde concentrations are used in vaccines.

Preserving Food

A. Label each of the following statements as true or false. If false, correct it.

1. Psychrophilic microorganisms can grow in refrigeration.
2. Most canning procedures do not attempt to remove the endospores of *Clostridium*.
3. Pasteurization is a special chemical treatment of milk.
4. Pasteur developed the process of pasteurization to kill lactic acid bacteria in wine to prevent it from spoiling.
5. A low pH is an effective means of chemical control against many microorganisms.
6. Salting is a physical control method used most often for fruit.

Multiple Choice: Review

1. Select the term that involves a treatment to destroy all microbial life.

   A. antisepsis
   B. decontamination
   C. disinfection
   D. sterilization

2. The D-value is the time in minutes required to kill _____ percent of the cells in a population.

   A. 90
   B. 75
   C. 50
   D. 10

3. Refrigerators are usually set at _____ degrees Celsius.

   A. 0
   B. 2
   C. 5
   D. 15

4. Moist heat treatment kills cells by

   A. alternating osmotic conditions.
   B. blocking UV light.
   C. dehydrating cells.
   D. denaturing proteins..

5. The TDP is the _____ temperature required to kill all microorganisms in a particular suspension in ten minutes.

   A. highest
   B. lowest

6. An autoclave normally maintains a temperature of _____ degrees Celsius at 15 pounds per square inch.

   A. 85
   B. 100
   C. 121
   D. 155

7. Which kind of bacterium can grow in a refrigerator?

   A. barophile
   B. mesophile
   C. psychrophile
   D. thermophile

8. The most lethal wavelength of UV radiation to microorganisms is at _____ nm.

   A. 50
   B. 150
   C. 265
   D. 420

9. A _____ percent solution of hydrogen peroxide is used as a weak antiseptic for cleaning wounds.
    A. 1
    B. 3
    C. 12
    D. 15

10. Select the incorrect association.
    A. alcohol/ethanol
    B. alkylating agent/ethylene oxide
    C. enzyme/perioxidase
    D. halogen/hydrogen peroxide

Answers
Correlation Questions

1. The population is reduced in size to ten percent of the original size through each cycle as it is eradicated. There are six cycles. The pattern of size reduction is: 100,000 – 10,000 – 1000 – 100 – 10 – 1 – 0. Two hours (120 minutes) divided by 6 cycles equals a D value of about 20 minutes.

2. Calculus includes the patterns of exponential growth that occur during the log phase. There are many patterns of statistical analysis applied to bacterial populations.

3. The salt established high osmotic strength outside the cells. The heat and UV light from the sun are sterilizing agents.

4. The death of a bacterial population refers to eradicating the entire population rather than killing one multicellular organism.

The Way Microorganisms Die

A. 1. sterilization   2. antisepsis   3. inhibits rather than kills   4. ninety   5. log   6. *Clostridium botulinum*
    7. lowest   8. minimum

Physical Controls on Microorganisms

A. 1. False; UV light only affects the surface.
    2. True
    3. True
    4. False; A low temperature slows down metabolism of most microorganisms but usually does not kill them.
    5. False; Short wavelengths of light can kill microbes.
    6. True
    7. False; X rays remove electrons from atoms.

8. True
9. False; Viruses are too small to be removed by the filters.
10. False; Sublimination is the direct conversion from a solid to a gas.

Chemical Controls on Microorganisms

A. 1. kills 2. germistat 3. 1200 4 coefficient 5. one thousand 6. inhibition 7. membrane
   8. hydroxyl 9. alcohol 10. three to six

B. 1. D 2. C 3. D 4. B

Preserving Food

A. 1. True
   2. False; Canning usually does remove the endospores of Clostridium botulinum.
   3. False; Pasteurization is the heating of milk for a length of time to remove microorganisms.
   4. True
   5. True
   6. False; Salting is usually employed to preserve fish or meat.

Multiple Choice: Review

1. D 2. A 3. C 4. D 5. B 6. C 7. C 8. C 9. B 10. D

Chapter 10
Classification

Principles of Biological Classification
    Scientific Nomenclature
    Artificial and Natural Systems of Classification
    The Fossil Record
    The Concept of Species
Microorganisms and Higher Levels of Classification
Methods of Microbial Classification
    Numerical Taxonomy
    Traditional Characters Used to Classify Prokaryotes
    Comparing Genomes
    Characters Used to Classify Viruses
    Dichotomous Keys
Summary

Key Terms

hierarchy
taxonomy
taxon
genus
specific epithet
phylogenetic
fossil
stromatolite
microbial mats
strain
protists
Eucarya
Bacteria
Archaea

characters
numerical taxonomy
dendogram
morphology
biochemistry
physiology
serology
phage typing
host range
nosocomial infection
sequencing DNA
DNA hybridization
probe
dichotomous key

Study Tips

1. Answer the several kinds of questions at the end of this chapter. Ask your instructor to check your results. Discuss your answers with the other students in the class.

2. Can you make a dichotomous key? A simple example is sufficient to help you understand the concept of keying. Use a series of geometric shapes for this process: circle, square, rectangle, triangle, hexagon. What character can you use for a first pair of steps that will broadly separate these shapes?

3. Study the protozoa in Chapter 12. Construct a dichotomous key to sort out the major groups based on their differences in motility. Write at least two pairs of statements in your key.

4. How are you progressing in lab? Ask your instructor if you will be identifying an unknown bacterium in this part of the course. Can you list all the Gram-positive and Gram-negative bacteria you have studied to date? Can you use this information as a good first step at keying out an unknown bacterium?

5. What other tests can you use to identify and classify a bacterial unknown? Are there other kinds of stains? Are metabolic tests (e.g., starch hydrolysis, fermentation of glucose) also applicable? Can you further refine the key you started in #3?

Correlation Questions

1. How do you think evolution has produced the different subspecies of humans throughout the geographical regions of the world, based on differences in mutations and natural selection? What do you think the future evolution of humans will do to the establishment of these subspecies?

2. New variations of bacteria are discovered on a planet through space exploration. These variations differ by their means of motility. What are the strengths and weaknesses for using motility as a basis for classification?

3. The address at which you live is an example of a hierarchical scheme of classification. Explain.

4. A fossil is discovered in a geological structure in Canada. The carbon in the rocks surrounding the fossil has a half-life of 2,000,000 years. In that time, one-half of the carbon changes into nitrogen-14. In the rocks with the fossil, one-fourth of the carbon/nitrogen is carbon is C-12 and the other three-fourths is N-14. What is the approximate age of the fossil? Consult the Internet to learn more about carbon dating.

Principles of Biological Classification

A. Complete each of the following statements with the correct term.
   1. From species to kingdom, a hierarchical scheme of classification ranks groups that are progressively broader and more _____.
   2. _____ is the science of classification.
   3. The _____ is the most exact taxon where all members are very similar.
   4. In *Staphylococcus aureus*, *aureus* is the _____ name.
   5. *E. coli* lives normally in the _____ of the human body.
   6. A classification based on phylogeny is a _____ scheme rather than an artificial scheme.
   7. The color of the colonies of *Staphylococcus aureus* is _____.
   8. _____ are the fossilized remains of phototrophic prokaryotes.
   9. The fossil record is limited to the past __(number)__ million years.
   10. Groups of bacteria do not always reproduce sexually and share in a common gene _____.
   11. The species of a bacterium can be subdivided into _____.
   12. The broadest taxon for classifying viruses is the _____.

B. Label each one of the following statements as true or false. If false, correct it.
1. The order is broader and more inclusive as a taxon than the genus.
2. *E. coli* is named in part by where it lives.
3. *Staphylococcus aureus* is named by the arrangement and color of its cells.
4. The Linnaean system of classification is natural.
5. The fossil record is extensive for the last 1.8 million years.
6. Fungi have prokaryotic cells.
7. The strains of *E. coli* are different species of bacteria.
8. Viruses are not classified into kingdoms.

Microorganisns and Higher Levels of Classification

A. Short Answer
1. Why did Haeckel propose a third kingdom, the protists?
2. What type of classification scheme for kingdoms did Whitaker propose?
3. How do the species in Archaea differ from the other bacteria?
4. Why was the Linnaean scheme of two kingdoms found unsatisfactory by most scientists?
5. What are the phyla of eukaryotic organisms?

Methods of Microbial Classification

A. Short Answer
1. What does a high similarity coefficient indicate when classifying microorganisms?
2. What kind of characteristic does the Gram stain provide for bacterial classification?
3. A suspension of bacterial cells is added to tetrazolium dye with a carbon source. The dye changes color. What does this indicate?
4. What kind of molecule is an antibody?
5. How are viruses used in phage typing?
6. If 300 G-C base pairs out of 800 base pairs in a DNA sample, the percentage of A-T base pairs is __(number)__.
7. How is DNA hybridization used for classification?
8. How is the sequencing of DNA used for classification?

Multiple Choice: Review

1. Select the most specific, least inclusive taxon from the following list.
   A. family
   B. genus
   C. kingdom
   D. order

2. In the binomial name, *Pseudomonas aeruginosa*, the first name indicates the

   A. family.
   B. genus.
   C. order.
   D. species.

3. The fossil record is limited to about the last _____ million years.

   A. 200
   B. 400
   C. 600
   D. 800

4. Stromatolites are

   A. bacterial life cycles.
   B. competing bacterial imprints.
   C. finely layered rocks.
   D. taxonomic categories.

5. In early classification schemes bacteria were classified as

   A. animals.
   B. fungi.
   C. plants.
   D. protozoa.

6. In early classification schemes protozoans were classified as

   A. animals.
   B. bacteria.
   C. fungi.
   D. plants.

7. The broadest taxonomic category used to classify viruses is the

   A. family.
   B. kingdom.
   C. order.
   D. phylum.

8. Which one of the following is mainly a biochemical and physiological characteristic?

   A. cell shape
   B. DNA sequence
   C. habitat preference
   D. pH growth requirement

9. Phage typing uses

   A. bacteria to classify viruses.
   B. viruses to classify bacteria.

10. A DNA sample has 2000 base pairs. 1100 of these base pairs are G-C. The percentage of A-T base pairs is
    A. 40
    B. 45
    C. 55
    D. 65

Answers

Correlation Questions

1. Each geographical population had its own unique group of mutations (e.g., eye color) and environmental selection pressures (e.g., climate). This explains the partial speciation leading to the subspecies or races of humans. In the future the elimination of geographical boundaries with improved transportation and communication will probably lead to the sharing of the collective, geographical gene pools and the loss of these individual subspecies.

2. Classification based merely on motility differences is superficial unless it can also be correlated with phylogenetically-based similarities and differences.

3. Consider your address: planet – continent – country – state – county – town – street as one example of a hierarchy from the broad and inclusive to the more exact and specific.

4. One-half of the C-14 changes to N-14 in 2 million years, establishing a 50% - 50% ratio of these two elements. In another 2 million years, one-half of the remaining C-14 changes into N-14, producing a ratio of one-fourth C-14 and three-fourths N-14. This requires a total of 4 million years in the rocks where the fossil was found. Therefore the age of the fossil is about 4 million years.

Principles of Biological Classification

A. 1. inclusive 2. taxonomy 3. species 4. species 5. colon 6. natural 7. yellow 8. stromatolites
   9. 600,000 10. pool 11. strains 12. family

B. 1. True
   2. True
   3. True
   4. False; It is artificial, based on visible similarities.
   5. False; An extensive fossil record exists for the last 600,000 million years.
   6. False; Fungi are eukaryotic.
   7. False; The strains of *E. coli* are genetic variations of the same species.
   8. True

Microorganisms and Higher Levels of Classification

A. 1. Microorganisms cannot be logically classified as plants or animals.

    2. He proposed a five-kingdom scheme. In addition to plants, animals, and protists, he added the kingdoms of monerans (prokaryotes) and fungi.

    3. The bacteria of Archaea are ancient bacteria and are unrelated to the true bacteria.

    4. Some species, such as bacteria and protozoa, are not very similar nor phylogenetically related to plants and animals.

    5. The phyla of eukaryotes are the animals, plants, fungi and protists.

Methods of Microbial Classification

A. 1. The organisms are closely related at the species or genus level.

    2. The Gram stain reveals mainly a morphological characteristic.

    3. The dye is reduced and the bacterium can use the carbon source.

    4. It is a protein molecule.

    5. Phages are used to classify bacteria. If a particular virus lyses a bacterium, there is a clear zone around the bacterium on the lawn of an agar surface. This is a distinctive clue for classification.

    6. 500/800 or 62.5%

    7. The more hybridization between DNA strands from two organisms, the closer they are related.

    8. The greater the similarity between the DNA sequences of two organisms, the more closely they are related.

Multiple Choice: Review

1. B   2. B   3. C   4. C   5. C   6. A   7. A   8. D   9. B   10. B

Chapter 11
The Prokaryotes

Prokaryotic Taxonomy
The Bergey's Manual Scheme of Bacterial Taxonomy
Domain Archaea
    A1 Crenarchaeota
    A2 Euryarchaeota
Domain Bacteria
    B4 Deinococcus-Thermus
    B5 Chrysiogenetes
    B10 Cyanobacteria
    B11 Chlorobi
    B12 Proteobacteria
    B13 Firmicutes
    B14 Actinobacteria
    B16 Chlamydiae
    B17 Spirochaetes
    B20 Bacterioidetes
    B21 Fusobacteria
Summary

Key Terms

methanogens
halophilic
bacteriorhodopson
thermoacidophile
Taq polymerase
heterocyst
crown
opine
transgenic plant
prosthecate
meninges
sheath
enteric bacteria
butanediol

acetoin
luminescent
xanthan gum
gliding
diplococci
aerotolerant anaerobe
obligate fermenter
snapping post-fission movememt
mycolic acid
leprosy
actinomycetes
multiocular
purple nonsulfur bacteria

Study Tips

1. Answer the several kinds of questions at the end of the chapter. After answering them, discuss the answers with your instructor and classmates.

2. This chapter uses a phylogenetic approach to classify bacteria. Can you organize the information in the chapter several ways to help you understand the facts and concepts? Here are some ideas.

Can you list the prominent Gram-positive and Gram-negative organisms described in the chapter? Are there other ways to group the bacteria?

3. Ask your instructor to provide some lists of bacteria for keying. Make a dichotomous key for each list. This process will help you for the identification of the bacterial unknown in lab.

4. Can you tie in the examples from this chapter and lecture with your lab experiences? As you observe them under the microscope, which bacteria have independent motility? Which ones are Gram-positive? Which ones are rod-shaped? Which ones have chains of spherical cells? Which ones can ferment lactose?

Correlation Questions

1. Is it logical to classify bacteria based on differences in their Gram-stain reactions?

2. What adaptations allow thermoacidophiles to tolerate their harsh environment?

3. How do you think luminescent bacteria can produce light without significant heat production?

4. How does the use of dry heat during cold weather make humans more susceptible to microbial infections?

Prokaryotic Taxonomy

A. Short Answer
   1. What is the recently-developed natural scheme used to classify bacteria?
   2. The most complete, widely accepted natural scheme of classification used the sequence of bases in _____.

The Bergey's Manual Scheme of Bacterial Taxonomy

A. Domain Archaea
   1. All bacteria of the phylum Crenarchaeota share the common property of being extremely _____.
   2. There are __(number)__ classes in the phylum Euryarchaeota.
   3. The methanogens make methane,known as _____.
   4. The halophilic archaea cannot grow with less than __(number)__ percent NaCl.
   5. The thermoacidophiles grow at temperatures near the _____ of water.

B. Domain Bacteria
   1. *Thermus aquaticus* produces the enzyme _____.
   2. Cyanobacteria are the main _____ fixers in nautre.
   3. Members of the phylum Chlorobi are _____; they must have light.
   4. Species of *Chlorobium* are the _____ bacteria.

5. *Acetobacter* produces _____ acid.
6. Members of *Agrobacterium* cause diseases in _____.
7. A _____ plant contain genes from another organism.
8. *Rhizobium* and *Bradyrhizobium* are symbiotic _____ fixing bacteria.
9. Bacteria with _____ have a spidery appearance.
10. Prosthecae increase the _____ of a cell.
11. *Axospirillum* is a _____ nitrogen-fixer.
12. Some members of Betaproteobacteria form _____, long tubes surrounding their cells.
13. Enteric bacteria infect hosts' _____ system.
14. All enteric bacteria have cells shaped similar to _____.
15. *E. coli* converts tryptophan to _____.
16. Some members of *Vibrio* have _____ (light-emitting) ability.
17. Some species of *Pseudomonas* are important to _____ ecology.
18. *Pseudomonas aeruginosa* can cause infections in victims suffering from _____.
19. *Xanthomonas* produces xanthum _____.
20. Some species of Deltaproteobacteria are _____ bacteria.
21. Deltaproteobacteria has species that participate in the cycling of _____ compounds.
22. The fruiting bodies of *Myxococcus* are _____-shaped.
23. Members of the genera _____ and _____ form endospores
24. A species of the genus _____ causes anthrax.
25. The species of the genus _____ causes tetanus.
26. A bacterial species of the genus _____ causes pneumonia.
27. *Bdellovibrio bacteriovorus* attacks other _____.
28. The term Mollicutes is synonymous with_____.
29. Rhinitis is an infection of the _____
30. *Actinobacter* has _____ movement.
31. The acid-fast genera are _____ and _____.
32. The actinomycetes are _____-forming bacteria.
33. The chlamydiae are Gram _____ bacteria.
34. Rickettsia reproduce by _____ division.
35. *Treponema pallidum* is a _____ by cell shape.
36. The Bacterioidetes are the _____ bacteria.

Multiple Choice Review

1. A natural classification scheme for bacteria relies mainly on

    A. cell morphology.
    B. DNA sequencing.
    C. Gram-staining
    D. motility studies.

2. Select the incorrect characteristic about spirochetes.

    A. Their axial filaments produce unusual motility.
    B. They are Gram-positive.
    C. They are long and helical.
    D. They move rapidly under the microscope.

3. *Rhizobium* is important for

    A. genetic engineering.
    B. nitrogen fixation.
    C. wine fermentation.
    D. yogurt production.

4. Species of *Salmonella*, *Shigella*, and *Yersinia* have the common characteristic of

    A. carrying out nitrogen fixation.
    B. fermenting wine.
    C. infecting respiratory systems.
    D. living enteric bacteria.

5. Vibrios are

    A. curved rods.
    B. rods.
    C. spheres.
    C. spirochetes.

6. Species of Chlamydiae are Gram-

    A. positive.
    B. negative.

7. *Bacillus* is a

    A. Gram-negative rod.
    B. Gram-negative sphere.
    C. Gram-positive rod.
    D. Gram-positive sphere.

8. The cyanobacteria were once called the
    A. blue-green algae.
    B. green-algae.
    C. red algae.
    D. yellow-brown algae.

9. Which genus has a species causing tuberculosis?
    A. *Bacillus*
    B. Escherichia
    C. Mycobacterium
    D. *Staphylococcus*

10. Each is a subgroup of the domain Archaea except the
    A. halophiles.
    B. methanogens.
    C. sulfur-producers.
    D. thermoacidophiles.

Answers
Correlation Questions

1. This scheme is not sound phylogenetically. It centers of a morphological trait, the makeup of the cell wall.

2. Their adaptations center of the bonding of their proteins at the secondary and tertiary level of structure. The proteins are stabilized at these levels and not denatured in harsh conditions.

3. A luciferase enzyme can convert substrate to products, releasing light but not heat.

4. Dry heat drys out the mucous membranes of the upper respiratory tract, hindering a major line of defense to ward off infection microbes.

Prokaryotic Taxonomy

A. 1. The scheme uses DNA sequencing. It measures the degree of relatedness of two organisms and emphasizes a phylogenetic basis.
   2. 16S ribosomal RNA.

The Bergey's Manual of Prokaryotic Taxonomy

A. 1. thermophilic 2. five 3. natural gas 4. ten 5. boiling point

B. 1. Taq polymerase 2. nitrogen 3. photosynthetic 4. green 5. acetic 6. plants 7. transgenic 8. nitrogen 9. prosthecae 10. surface area 11. free-living 12. sheaths 13. digestive 14. rods 15. indole 16. luminescent 17. soil 18. burn 19. gum 20. gliding 21. sulfur 22. dome

23. *Clostridium, Bacillus* 24. *Bacillus* 25. *Clostridium* 26. *Streptococcus* 27. bacteria
28. mycoplasmas 29. nose 30. snapping post-fission 31. *Mycobacterium, Nocardia* 32. spore
33. negative 34. simple 35. spirochete 36. purple nonsulfur

Multiple Choice: Review

1.B 2.B 3.B 4.D 5.A 6.B 7.C 8.A 9.C 10.C

Chapter 12
Eukaryotic Microorganisms, Helminths, and Arthropod Vectors

Eukaryotic Microorganisms
> Fungi
> Algae
> Lichens
> Protozoa
> The Slime Molds
> Relatedness of Eukaryotic Microbes
> Helminths
> Arthropod Vectors
> The Organisms
> Arthropods and Human Health
Summary

Key Terms

helminth

arthropod

heterotrophic

nonphototrophic

absorptive

saprophyte

mycelium

hypha

thallus

budding

sporangium

conidiophore

ascospore

basidiospore

dimorphism

phytoplankton

lichen

flagellate

ameoboid

pseudopod

ciliate

trophozoite

sporozoite

merozoite

plasmodium

scolex

proglottid

Study Tips

1. Continue to answer the several kinds of questions at the end of the chapter.

2. Have you developed a study technique that works best for you? Here is one suggestion. Begin by scanning the major ideas in this chapter. Can you compare the fungi, algae, lichens, and slime molds for similarities and differences?

3. Vocabulary remains a major challenge for any student to master in a science course. Begin by listing the key terms down the lefthand side on a sheet of paper. After studying the chapter, write a definition across from each term in your own words.

4. Up to this point, most of your study has been devoted to the bacteria and viruses. This chapter presents new groups of microorganisms. It also presents the helminths and arthropod vectors. Your lab experience can help extend your study of the unique groups of organisms discussed in this chapter.

5. Continue to consult the Internet to expand your knowledge on the subject of microbiology. Key search terms from this chapter include: saprophyte, helminth, protozoa, etc.

Correlation Questions

1. Consider classifying slime molds into their own unique kingdom. What unique characteristics does this group have to justify this new classification?

2. To some biology students, fungi appear to be plants rather than members of a separate kingdom. How do you think this mistake could be made?

3. A researcher discovers a new invertebrate with an exoskeleton. How can this biology proceed to learn about the chemical composition of this structure? How is this exoskeleton adaptive?

4. A new plantlike organism is discovered by a researcher. How can this biology find out what the highest level of organization (e.g., tissue, organ) is for this organism?

Eukayrotic Microorganisms

Fungi

A. Label each one of the following statements as true or false. If false, correct the statement.
1. A green organism is discovered in a forest ecosystem. Analysis of its cells reveals chloroplasts. This organism could be a fungus.
2. The kingdom of fungi does not contain any single-celled organisms.
3. A mycelium is a mass of hyphae.
4. The thallus is the body of a fungus.
5. An organism is observed in a controlled laboratory environment. It grows rapidly in low-oxygen conditions. There is a high probability that this organism is a fungus.
6. The fungi are a major group of decomposers in the environment.
7. Sporangiospores are produced in conidiophores.
8. The cell structure of a microorganism is observed under the microscope. The cells are well-defined and each cell has one nucleus. This organism could be a lower fungus.
9. The oomycetes cannot live in the soil, as water is required for their survival.
10. In the life cycle of *Rhizopus*, a diploid zygospore divides to produce a structure that develops into a sporangium.
11. The species of Deuteromycetes form basidiocarps.
12. The dikaryon is a mycelium with two different kinds of nucleii.

13. A microorganism is studied under the microscope. It consists of long chains of conidia. This microorganism could be a species of the genus *Penicillium*.

14. During the baking of bread, *Saccharomyces cerevisiae* makes the dough of the bread rise through the process of photosynthesis.

15. One species of yeast is a facultative anaerobe. Its growth can be inhibited by placing it in a high-oxygen environment.

16. Most plant diseases are caused by fungi.

17. A fungus caused Dutch elm disease.

18. Fungal infections are called mycoses.

19. Alfatoxin is a fungal toxin.

20. Headaches in humans are caused by a fungus.

Algae

A. Complete each of the following statements with the correct term or terms.

1. Phytoplankton use the gas _____ in the process of photosynthesis.

2. Algae lack the _____ structure of higher plants.

3. The colorless alga _____ can cause bursitis in humans.

4. From a red alga, the product _____ is used as a milk thickener.

5. _____ earth is used to filter liquids.

6. In the absence of the pigments _____ and _____, chlorophyll cannot receive light to conduct photosynthesis.

7. 80% of the atmosphere of the Earth is carried out by _____ in the ocean.

8. _____ is the study of algae.

9. Absence of an _____ in *Euglena* will inhibit its ability to detect light.

10. A microorganism cannot be a species of *Chlamydomonas* if its flagellum is on the _____ side of the body.

Lichens

A. Complete each of the following statements with the correct term or terms.

1. A lichen is a symbiotic association between a _____ and a _____.

2. A lichen is studied in lab. Most of its body mass consists of a _____ organism.

3. *Roccella tinctoria* is the source of an indicator that turns from blue to _____ in an acid environment.

4. A cyanobacterium in a lichen fixes _____, an element found in proteins but not found in sugars.

Chapter 12

Protozoa

A. Match each description to its correct group of protozoans.

1. move by a long, whiplike organelle
2. move by pseudopodia
3. *Plasmodium* is a member
4. all are parasites
5. move by short, numerous hairlike projections
6. *Paramecium* is a member
7. contains *Entamoeba*
8. species have a macronucleus and micronucleus
9. one species causes malaria
10. contains the trypanosomes
11. diplomonads are included
12. sporozoite infects red blood cells
13. merozoite is part of one species' life cycle
14. *Didinium* attacks a species of this group
15. *Giardia* is a member

A. Ciliophora (Ciliates)

B. Flagellates

C. Amoeboids

D. Sporozoa

Slime Molds

A. Complete each one of the following with the correct term.

1. Species of Myxogastria are the _____ slime molds.
2. A plasmodium is a _____ cytoplasm.
3. Spores of Myxogastria are _____, as their spores contain only one set of chromosomes.
4. Species of Acrasieae are the _____ slime molds.
5. A grex will probably show reduced motility in the absence of _____.

Helminths

A. Flatworms

1. Select the characterisitics that are descriptive of the body of species of Platyhelminthes.

   A. radial symmetry
   B. flattened body
   C. digestive tract present in all parasites
   D. nervous system present
   E. excretory system present

2. List five characteristics of the species of Cestoda.

3. What are cystecerci?
4. Outline the series of larvae of a fluke after it infects the lung of a human host.

B. Roundworms
   1. Select the characteristics that are not descriptive of the nematode body.
      A. bilateral symmetry
      B. flattened body
      C. complete digestive tract
      D. well-developed reproductive system
      E. have larval and adult forms
      F. larvae do not form cysts in muscles

Arthropod Vectors

A. Complete each of the following with the correct term or terms.
   1. What are some key adaptations accounting for the success of arthropods on the Earth?
   2. What is another name for the group of ticks and mites?
   3. Why is a study of the arthropods important in microbiology?
   4. Name the three major body regions of an insect?
   5. By number of species, how do you evaluate the success of arthropods?
   6. What is transovarial transmission?

Multiple Choice: Review

1. Select the correct description about the fungi.
   A. None of their species are parasites.
   B. Some form hyphae and mycelia.
   C. They are phototrophic organisms.
   D. They have a prokaryotic cell structure.

2. Among yeasts budding is a means of
   A. energy storage.
   B. feeding.
   C. movement
   D. reproduction.

3. Which class is not a member of the lower fungi?
   A. Ascomycetes
   B. Chytridiomycetes
   C. Oomycetes
   D. Zygomycetes

4. Select the correct statement about the Deuteromycetes.

   A. They are identified by their pigmentation system.
   B. They belong to the lower fungi.
   C. They produce sexual spores.
   D. They produce only conidia.

5. Phycology is the study of

   A. algae.
   B. bacteria.
   C. fungi.
   D. plants.

6. A lichen is an example of a symbiotic association called

   A. commensalism.
   B. mutualism.
   C. parasitism.
   D. predation.

7. Protozoa are classified into groups based on differences in their

   A. cell shape.
   B. means of motility.
   C. means of locomotion.
   D. type of reproduction.

8. An algae with flagella is the

   A. amoeba.
   B. didinium.
   C. euglena.
   D. paramecium.

9. Name the free-swimming larva of *Paragonimus westermani* that encysts in crayfish.

   A. cercaria
   B. cystecerci
   C. miracidia
   D. redia

10. A vector

   A. covers the surface of an arthropod body.
   B. is a specialized reproductive cell.
   C. serves as an antigenic substance.
   D. transports a microbe from one host to another.

Answers

Correlation Questions

1. Start with some of their plantlike characteristics. Compare these to their similarities to protozoans. Next, view their unique characteristics such as the grex and pigment systems.

2. Consider the leafy appearance of some along with the fact that they can grow off a substrate and do not move.

3. Test samples of the exoskeleton with different enzymes for polysaccharides. Use enzymes such as cellulase, amylase (for starch), enzymes of chitin, and other catalysts that can chemically digest polysaccharides. Continue with proteases to test for protein content to learn about the chemical content of this structure which is adaptive by the protection it offers.

4. Study the makeup of the cells through the microscope and by observing the responses these cells can make. Do the cells work independently or are the cell integrated, indicating a true tissue?

Eukayrotic Microorganisms

Fungi

A. 1. False; A fungus is saprophytic. It is not photosynthetic.
   2. False; Yeasts are single-celled.
   3. True
   4. True
   5. False; Most fungi are aerobes.
   6. True
   7. False; Sporangiospores are produced on sporangia.
   8. False; Lower fungi are coenocytic.
   9. False; They need water, but a film of water exists around soil particles.
   10. True
   11. False; Species of Basidiomycetes form basidiocarps.
   12. True
   13. True
   14. False; It produces carbon dioxide by fermentation.
   15. False; It can grow in the presence or absence of oxygen.
   16. True
   17. True
   18. True
   19. True
   20. False; Fungal products are used to treat headaches.

Algae

A. 1. carbon dioxide   2. tissue   3. *Prototheca*   4. carrageenan   5. diatomaceous
    6. carotenoids and phycobilins   7. phytoplankton   8. phycology   9. eyespot   10. anterior

Lichens

A. 1. fungus, phototroph   2. fungal   3. red   4. nitrogen

Protozoa

A. 1. B   2. C   3. D   4. D   5. A   6. A   7. C   8. A   9. D   10. B   11. B   12. D   13. D   14. A   15. B

The Slime Molds

A. 1. true   2. multinucleate   3. haploid   4. cellular   5. light

Helminths

A. 1. B, D, E
    2. flattened, segmented body; scolex, proglottids, germinal center, hermaphrodite
    3. They are the larvae of *Taenia saginata* embedded in the skeletal muscles of a host.
    4. miracidia - redia - cercaria

B. 1. B, F

Arthropod Vectors

A. 1. exoskeleton, three body segments, and jointed appendages
    2. arachnids
    3. They are vectors for the transmission of microorganisms.
    4. head, thorax, and abdomen
    5. There are over 800,000 species of insects alone. They have high species diversity. These various species occupy many different ecological niches.
    6. Ticks pass infectious microorganisms into their eggs.

Multiple Choice: Review

1. B   2. D   3. A   4. D   5. A   6. B   7. B   8. C   9. A   10. D

Chapter 13
The Viruses

Key Terms

virus	lysogenic pathway
virion	virulent
bacteriophage	temperate
host range	prophage
capsid	primary cell culture
envelope	diploid cell line
plus-strand	heteroploid cell line
minus-strand	continuous cell line
capsomere	maturation
icosohedral	release
polyhedral	retrovirus
adsorption	antigenic drift
penetration	reverse transcriptase
uncoating	benign

lawn	invasive
circular plaque	malignant
latent period	oncogene
burst period	TMV
burst size	mycovirus
eclipse period	viroid
lytic pathway	prion

Study Tips

1. Answer the several kinds of questions at the end of the chapter. Discuss the answers with your instructor and classmates.

2. From your study of the chapter and lecture notes, sketch the basic makeup of a virus. Check the text for accuracy after you have made these sketches on your own.

3. Outline the life cycle of the virulent phage and temperate phage. Express these outlines in your own terms rather than simply copying the information from the text. Ask your instructor to evaluate the accuracy of your outlines.

4. Viruses are constantly in the news. Select one recent article that your can find from an Internet search. Discuss how the following viral diseases impact human society: the common cold, Herpes, AIDS, and several forms of human cancer.

Correlation Questions

1. Antibiotics are usually not effective in treating viral diseases. Discuss the reasons for this based on viral structure.

2. Viruses have recently been used to treat human diseases. How are they used this way?.

3. Although the subjects of nomenclature and taxonomy are related, they are different. How?

4. How would you develop a scheme to classify viruses into a unique, seventh kingdom?

The Ultimate Parasites

A. Describe the contributions of each of the following scientists to the discovery of viruses.
   1. Iwanowski -
   2. Beijerinck -
   3. Stanley -

B. Summarize the characteristics of viruses.

Classification of Viruses

A. Label each of the following as true or false. If false, correct it.
   1. Bacteriophages are viruses.
   2. A specific virus usually attacks a variety of species.
   3. The T4 virus contains over 300 genes.
   4. The RNA molecules of RNA viruses never store genetic information.
   5. Minus strand RNA is converted into mRNA after it enters a host cell.
   6. The capsid is the nucleic acid core of the virus.
   7. The most common polyhedral virus is the icosahedral virus.
   8. The membrane that surrounds enveloped viruses is a piece of the host cell's membrane.
   9. Nonpolar solvents destroy the genes in a virus.
   10. By penetration a virus attaches to the host cell membrane.

B. Complete the following statements.
   1. The ICTV scheme for classifying viruses has __(number)__ taxonomic levels.
   2. The taxonomic levels for classifying viruses by the ICTV scheme are:

Bacteriophages

A. Complete each of the following statements with the correct term.
   1. The first bacteriophages studied were seven examples that attacked the organism _____.
   2. By the plaque count a nutrient agar surface first supports a _____ of host bacterial cells.
   3. By the plaque count a plaque is the result of an epidemic started by one _____.
   4. In the one-step growth curve there is no viral increase during the _____ period.
   5. The _____ size is the average number of new virions released from a cell infected by viruses.
   6. _____ phages follow only the lytic pathway.
   7. _____ phages follow the lytic or lysogenic pathway.
   8. _____ is when the component parts of new virions are assembled near the end of the latent period.
   9. The phage DNA integrated into a host chromosome is a _____.
   10. Only _____ strains of *C. diphtheriae* cause diptheria.

Animal Viruses

A. Complete each of the following statements with the correct term.
   1. In the 1930s virologists learned that some viruses could be cultured in embryonated _____ eggs.
   2. A _____ is a confluent of animal cells, one layer thick, in a petri dish.

3. _____ cell cultures are started from normal tissues taken directly from humans or other animals.

4. Continuous cell lines are usually derived from _____ tissue.

5. The _____ cell line is the most famous continuous cell line.

B. Describe the events of each of the following stages of viral replication.

1. adsorption -

2. penetration -

3. uncoating -

4. viral synthesis -

5. maturation -

6. release -

7. latency -

C. Match each of the descriptions to the correct animal virus.

1. families of A, B, C   A. retrovirus

2. uses a reverse transcriptase   B. influenza virus

3. has NA spikes   C. tumor virus

4. some cause malignant growths

5. RNA is eight separate pieces

6. the capsid is truncated

7. plus-stranded RNA virus

8. has HA spikes

Plant Viruses

A. Label each of the statements as true or false. If false, correct it.

1. Plant viruses are usually named by their host range.

2. Aphids transmit plant viruses.

3. Plant viruses are never enveloped.

4. Most viral diseases of plants quickly kill the plants.

5. TMV is named by it unique protein coat.

Viruses of Eukaryotic Microorganisms/Infectious Agents That Are Simpler Than Viruses

A. Complete each of the following statements with the correct term.

1. Until __(number)__ years ago, microbiologists did not believe that viruses infected eukaryotic cells.

2. _____ are viruses that infect fungi.

3. A _____ is a circular molecule of ssRNA without a capsid.

4. The spindle tuber viroid is composed of only __(number)__ nucleotides.
5. The term _____ means proteinaceous infectious particles.

The Origin of Viruses

A. 1. Explain an accepted theory among scientists about how viruses evolved.

Multiple Choice: Review

1. Select the incorrect statement about viruses.
   A. All cellular organisms are attacked by them.
   B. They are parasites.
   C. They contain a nucleic acid inside a protein coat.
   D. They have a cellular makeup.

2. Viruses are classified by each of the following except
   A. host range.
   B. life cycle.
   C. lipid synthesis.
   D. size.

3. The viral nucleic acid enters the host cell. This step of its life cycle is
   A. adsorption.
   B. bursting.
   C. penetration.
   D. release.

4. The broadest taxonomic category applied to viruses is the
   A. class.
   B. family.
   C. kingdom.
   D. phylum.

5. PFUs are
   A. infected cells only.
   B. virions only.
   C. infected cells and virions.
   D. neither infected cells nor virions.

6. The first stage of viral replication is
   A. adsorption.
   B. maturation.
   C. penetration.
   D. release.

7. Virulent phage infections are swift and deadly.

   A. True
   B. False

8. Phages contain each of the following kinds of RNA except

   A. dsDNA
   B. dsRNA
   C. ssRNA
   D. tsRNA

9. Plant viruses are normally named by the

   A. amount of pigmentation.
   B. disease they cause.
   C. host range.
   D. size of their particles.

10. HIV/AIDS is a

   A. influenza virus.
   B. retrovirus.
   C. TMV.
   D. tumor virus.

Answers
Correlation Questions

1. Antibiotics usually attack a bacterial cell wall or ribosomes. These structures, along with other organelles (e.g., mitochondria) are absent in viruses. As parasites, viruses invade host cells and take over the metabolism of the host cells, using their mitochondria to make ATP and using their ribosomes to make proteins. This is necessary to complete their life cycle as they reproduce at the expense of the host cell.

2. These viruses are serving as vectors, carrying DNA into host cells to replace the defective DNA (e.g., the gene for cystic fibrosis) with genetic material that will not cause diseases..

3. Taxonomy is the meaningful classification of organism into categories. Nomenclature is applying a scientific name used to identify the organism that occupies these categories. The scientific name E. coli is its binomial name for its genus and species taxons.

4. This question has several possible answers. Start with the strategies of using the biochemical makeup of the virus: its nucleic acid core and its protein capsid. From a phylogenetic standpoint, relate this to the kind of eukaryotic cells they infected in their ancient past.

The Ultimate Parasites

A. 1. He found that filters that normally removed most microorganisms did not trap the causative agent in the extracts of plants with the tobacco mosaic disease.

2. He reported that microorganisms causing the tobacco mosaic disease are smaller than bacteria and must pass through the pores of the filters.

3. He crystallized the TMV virus, proving its existence.

B. 1. Viruses are intracellular parasites. They consist of a protein coat surrounding a nucleic acid. They can be viewed as bits of genetic information that instruct a host cell to make more viruses. They are small enough to pass through bacteria-proof filters. Some cause disease. Viruses lack a cellular structure.

Classification of Viruses

A. 1. True

2. False; A specific virus usually attacks only one species.

3. False; The T4 virus contains 77 genes.

4. True

5. True

6. False; The capsid is the protein envelope of the virus.

7. True

8. True

9. False; Nonpolar solvents destroy the viral cell membrane.

10. False; By adsorption a virus attaches to a host cell membrane. By penetration its nucleic acid enters the host cell.

B. 1. three

2. family, genus, and species

Bacteriophages

A. 1. *E. coli* 2. lawn 3. virion 4. latent 5. burst 6. virulent 7. temperate 8. maturation
9. prophage 10. lysogenic

Animal Viruses

A. 1. chicken 2. monolayer 3. primary 4. cancerous 5. HeLa

B. 1. Receptor-binding proteins of the virus bind to cell receptors of the animal cell.

2. Viruses enter host cells by fusing with the plasma membrane of the host cell, by phagocytosis, or by passing through the host cell membrane.

3. The capsid is removed, making the viral genes naked in the animal cells.

4. The viral nucleic acid takes over metabolism of the host cell, using the ribosomes, ATP, and other raw materials to make new virions.

5. Capsids assemble around the nucleic acids of the new virions.

6. The new virions push through the plasma membrane of the host cell and leave the cell.

7. The viral nucleic acid is incorporated into the host chromosome, but does not express itself. It remains latent. Later the virus may be activated by stimulus.

C. 1. B 2. A 3. B 4. C 5. B 6. A 7. A 8. B

Plant Viruses

A. 1. False; They are named by the disease they cause.
   2. True
   3. True
   4. False; They produce slow, degenerative diseases.
   5. False; It is named by the spotting pattern it causes on disease.

Viruses of Eukaryotic Microorganisms/Infectious Agents That Are Simpler Than Viruses

A. 1. forty 2. mycoviruses 3. viroid 4. 359 5. prion

The Origin of Viruses

A. 1. Viruses may have evolved from the cells they originally infected. They left as part of the genome of ancient cells and currently are infecting new cells.

Multiple Choice: Review

1. D 2. C 3. C 4. B 5. C 6. A 7. A 8. D 9. B 10. B

Chapter 14
Microorganisms and Human Health

Key Terms

normal biota
resident biota
transient biota
broad-spectrum antibiotic
opportunist pathogen
changing biota
bifidus factor
symbiosis
commensalism
mutualism
microbial antagonism
parasitism
pathogen

nonspecific surface defenses
mucous membrane
mucociliary system
lysozyme
adhesins
receptors
bacteriocidin
sebaceous
acne
axillary
blepharitis
meconium
peristalsis

Study Tips

1. Have you completed a course on the anatomy and physiology of the human body? Review the structure and function of the skin and the various tracts of the body: respiratory, digestive, urinary, and reproductive. This knowledge will help you understand the environment of the normal biota described in this chapter.

2. A review of cell structure will also help you study this chapter. For example, how do some pathogens overcome the mucociliary system of the respiratory tract?

3. There are many applications of chemistry to the topic of microorganisms and human health. For example, how does a change in the pH of the small intestine (pH of 8 to 9) change the normal biota of this body region?

4. Compose a table to summarize the normal biota of the various regions of the body. List the body regions along the lefthand side of the table. Across from each region, describe the normal biota.

5. Search the Internet to seek other sources of information on the topics of this chapter. Key search terms include: pathogen, mucous membrane, and peristalsis.

Correlation Questions

1. Why are space travelers checked for their microbiota upon return from their space travel?

2. The skin is a true organ. How can this be proven by studying its histology or tissue structure?

3. Compared to the term *microflora*, how is the term *microbiota* more accurate for describing the normal microbial inhabitants of the human body?

4. Cystic fibrosis inhibits the mucociliary action in the respiratory tract. How does this make the body more susceptible to infections?

Normal Biota

A. Define each of the following and provide an example of a microorganism in the human body for each.
   1. resident biota
   2. transient biota
   3. opportunist
   4. changing biota

B. Label each of the following body regions normally occupied by a normal biota population.

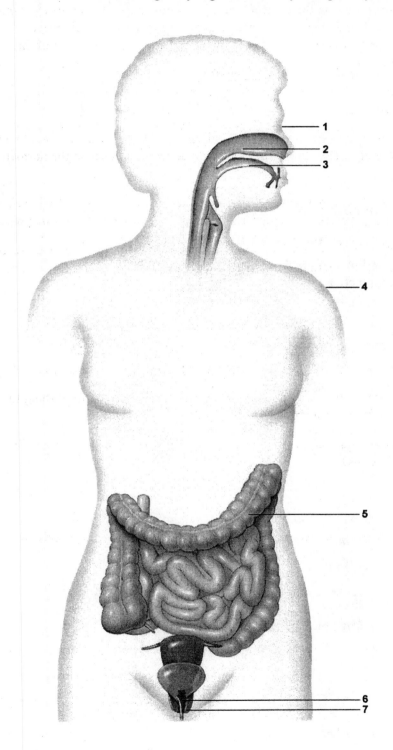

Symbioses

A. Match each of the following descriptions to the correct kind of symbiosis.

1. Both partners benefit.

2. The host is harmed.

3. One partner benefits and the other is neither helped nor harmed

4. The majority of human normal biota are involved in this type of symbiosis.

5. This type is lacking in humans.

6. One member is a pathogen.

A. commensalism.

B. mutualism

C. parasitism

Factors that Determine the Normal Biota

A. Label each of the following statements as true or false. If false, correct it.

1. Epithelial surfaces are a biochemical defense against microorganism invasion.

2. Compared to the skin, the conjunctiva is a thin membrane.

3. A mucous membrane covers the surface of the eye.

4. The mucociliary system is a biochemical defense mechanism..

5. The rapid flow of urine discharges bacteria from the urethra.

6. Keratin is a component of the inner lining of the digestive tract.

7. Bile is normally secreted into the stomach.

8. Lysozyme is normally secreted into the stomach.

9. Gram-positive bacteria are particularly vulnerable to the effects of lysozyme.

10. Members of the normal biota of epithelial surfaces attach to the epithelium by their pili.

Sites of Normal Biota

A. Select the correct statement from each of the following pairs of statements.

1. A. The dryness of the skin limits microbial growth.

   B. The skin makes up less than five percent of the body weight.

2. A. The skin secretes fatty acids.

   B. The skin does not secrete lysozyme.

3. A. The majority of the skin's normal biota are commensals.

   B. Fungi are not normal biota of the skin.

4. A. *Staphylococcus aureus* is found on everyone's skin.

   B. *Staphylococcus epidermidis* is found on everyone's skin.

5. A. Acne is caused by a virus.

   B. Aerobic diphtheroids grow near the surface of the skin.

6. A. Axillary normal biota live in the groin of a person.

   B. The conjunctiva is a mucous membrane.

7. A. Numerous microorganisms are inhaled daily.

   B. Cystic fibrosis patients produce a very thin mucus.

8. A. *Staphylococcus epidermidis* is found in the nasal cavity of most humans.

   B. *Moraxella catarrhalis* is a Gram-positive pathogen.

9. A. The mouth is an inhospitable environment for normal biota.

   B. The saliva contains lysozyme.

10. A. *Staphylococcus* species are widespread in the oral cavity.

    B. *Streptococcus* species are not found in the oral cavity..

11. A. Meconium lacks microorganisms.

    B. The pH of the stomach is very acidic.

12. A. Bile salts destroy many microorganisms in the intestine.

    B. Peristalsis is the chemical digestion of food.

13. A. The large intestine has a large, diverse normal biota.

    B. *Clostridium* is normally absent in the large intestine.

14. A. The vagina does not sustain a normal biota population.

    B. The vagina has a low pH.

15. A. The upper urethra contains many microorganisms.

    B. The urethral lining has tightly joined epithelial cells.

Is the Normal Biota Helpful or Harmful?

A. 1. State one reason that normal biota can be helpful to the body.

   2. State one reason that normal biota can be harmful to the body.

Multiple Choice: Review

1. Select the human body structure that has a normal biota population.

   A. blood
   B. lower respiratory tract
   C. skin
   D. skeletal muscles

2. Opportunists are microorganisms that

   A. always cause a disease.
   B. cause disease when the proper condition arises.
   C. constantly produce beneficial effect in the host.
   D. never cause disease.

3. Which type of symbiosis is most common between normal biota and humans?

    A. amensalism
    B. commensalism
    C. mutualism
    D. parasitism

4. The mucociliary system of the human body is a _____ defense.

    A. biochemical
    B. mechanical
    C. structural
    D. transport

5. Adhesin helps normal flora attach to _____ surfaces.

    A. connective tissue
    B. epithelial
    C. muscular
    D. nerve

6. Select the correct association.

    A. axillary/blood
    B. diphtheroids/present of all people's skin
    C. *Propionibacterium*/species causes acne
    D. *Staphylococcus epidermidis*/found on all people's skin

7. The conjunctiva is a covering of the

    A. brain.
    B. eye.
    C. heart.
    D. kidney.

8. *Candida albicans* is a

    A. bacterium living in the oral cavity.
    B. bacterium living in the upper respiratory tract.
    C. yeast living in the oral cavity.
    D. yeast living in the upper respiratory tract.

9. Select the bacterium that is not found in substantial numbers in the large intestine.

    A. *Bacteroides*
    B. *Bifidobacterium*
    C. *Escherichia*
    D. *Staphylococcus*

10. Microbially produced carcinogens are a main cause of cancer in humans.

    A. True
    B. False

Answers
Correlation Questions

1. Unknown species of microorganisms could have infected the travelers during their time in space. Humans may not have evolved immune mechanisms to defend against these microbes.

2. An organ is two or more tissues working together. The epidermis consists of epithelium. The underlying dermis is connective tissue.

3. *Microbiota* is a more inclusive term for all kinds of microorganisms. The term *microflora* has plantlike connotations.

4. The mucous membranes and ciliary action of the respiratory tract are a major line of defense protecting the body. Loss of their mechanisms leaves the body vulnerable to microbial attack.

Normal Biota

A. 1. They are characteristic and permanent biota in a given body region. One example is *E. coli* living in the intestinal tract.
    2. These biota cannot persist indefinitely in the body. Pathogenic strains of *Staphylococcus aureus* among hospital workers is one example.
    3. These microorganisms normally do not inhabit the body. However, they will cause a disease in the body when the proper opportunity arises.
    4. Some biota populations change in a body region over time. For example, intestinal biota change over the human lifespan.

B. 1. conjunctiva 2. nasal cavity and nasopharynx 3. mouth 4. skin 5. intestinal tract
    6. urethra 7. vagina

Symbioses

A. 1. B 2. C 3. A 4. A 5. B 6. C

Factors that Determine the Normal Biota

A. 1. False; They are a structural defense.
    2. True
    3. False; They are dead and are replaced by the division of cells in the deeper layers of the skin.
    4. False; It is a mechanism of the respiratory tract.
    5. True
    6. False; Keratin is a component of the outer layer of the skin.
    7. False; Bile is secreted into the small intestine.
    8. False; The stomach secretes HCl.
    9. True
    10. True

Sites of Normal Biota

A. 1. A 2. A 3. A 4. B 5. B 6. B 7. A 8. A 9. B 10. A 11. B 12. A 13. A 14. B 15. B

Is the Normal Biota Helpful or Harmful?

A. 1. A commensal can prevent a pathogen from establishing an environment in the body through microbial antagonism.

2. A commensal can become an opportunist and cause an infection in the body.

Multiple Choice: Review

1. C 2. B 3. B 4. B 5. B 6. C 7. B 8. C 9. D 10. B

Chapter 15
Microorganisms and Human Disease

Key Terms

virulence factor

reservoir

transmission

portal of entry

fomite

sexually transmissible disease

vertical transmission

prenatal

perinatal

airborne transmission

parenteral transmission

mechanical vector

biological vector

airborne transmission

filamentous hemagglutinin

avirulent

noninvasive

antigenic variation

IgA protease

serum resistance

transferrin

lactoferrin

ferritin

siderophore

pathogenesis

toxin

hypersensitivity

exotoxin

endotoxin

proteolytic toxin

adenyl cyclase system

antitoxin

toxoid

botulism

endocytosis

phagocyte

intracellular pathogen

pathogenicity

interleukin-1

cytolysin

autolysis

Study Tips

1. Interview a local ecologist or environmental biologist at your college. A brief visit with this person can provide you with additional insights about the concepts of this chapter.

2. Does your college have courses in the anatomy and physiology of the human body? Ask to arrange some study time in the lab for these courses. In the lab study the human torso model. Locate the different portals of entrance and exit described in this chapter.

3. Continue to answer the several kinds of review questions stated at the end of the chapter. Ask your instructor to provide answer keys to check the accuracy of your answers.

4. Continue to study vocabulary by writing out a list of key terms from the chapter. Define these terms in your own words.

5. Continue to use the Internet to expand your knowledge of the topics of microbiology. Use the Key Terms from the chapter as search terms.

Correlation Questions

1. A microbiologist wants to distinguish among the chronic carriers and incubatory carriers for a disease. How can he make this distinction?

2. A scientist wants to eradicate a disease which has a deer population as its reservoir. Concentrating on this deer population, how can the scientist begin to eradicate the disease?

3. A medical doctor vaccinates a person against a viral disease by using an injection with a hypodermic needle. How deep must this injection penetrate the skin? Why is this a major consideration?

4. A pathogen attacks a person, depleting her concentration of transferrin and ferritin. From this infection, the person develops anemia. Explain.

The Seven Capabilities of a Pathogen
One: Maintaining a Reservoir

A. Define each of the following.

1. virulence factor -

2. incubatory carrier -

3. chronic carrier -

4. zoonosis -

5. animal reservoir -

6. environmental reservoir -

Two: Getting To and Entering a Host

A. Label each of the following statements as true or false. If false, correct it.

1. Transmission refers to a pathogen leaving its reservoir and entering the body of a host.

2. The phrase "infectious dose" refers to the number of microbial cells that must enter the host body to cause death in 50 percent of the test animals.

3. The phrase "lethal dose" refers to the number of microbial cells that must enter the host body to cause infection in 50 percent of the test animals.

4. *Bordetella pertussis* is transmitted from host to host by respiratory droplets.

5. Whooping cough is a communicable disease.

6. A fomite is a microorganism causing a disease.

7. The most effective approach to decrease the spread of colds is through frequent handwashing.

8. STDs are spread when the cutaneous membrane of an infected person contacts the cutaneous membrane of a noninfected person.

9. Herpes simplex I is caused by a bacterium.

10. A perinatal infection occurs across the placenta, before birth.

11. Some arthropod vectors are mechanical vectors.

12. Malaria is caused by a biological vector.

13. Airborne diseases must be spread by microorganisms constantly located in the air.

14. A species of *Mycobacterium* causes tuberculosis.

15. Parenteral transmission occurs by an infection through the digestive tract.

Three: Adhering to a Body Surface

A. Complete each of the following statements with the correct term.

1. Pathogens attach to body surfaces by structures called _____.

2. The cells or tissues that a pathogen attacks are the pathogen's tissue _____.

3. *Bordetella pertussis* cells attach to the _____ of the epithelial cells of the respiratory tract.

4. FHA is a wavy _____ and not a true pilus.

5. *Neisseria gonorrhoeae* binds to the genital, _____ pharyngeal, and _____ surfaces of the host body.

Four: Invading the Body

A. Short Answer

1. Name two noninvasive pathogens.

2. How do invasive pathogens enter cells of the host?

3. What is the normal function of phagocytes?

4. Describe the key to success of *Coxiella burnettii* to cause human infection.

5. What are invasins?

6. What is the unique property of intracellular pathogens?

Five: Evading the Body's Defenses

A. Select the correct statement from each of the following pairs.

1. A. *Streptococcus pneumoniae* is a capsule-dependent pathogen.

   B. Most bacteria attacking the nervous system lack a capsule.

2. A. *Streptotoccus pyrogens* produces an IgA protease.

   B. *Neisseria gonorrhoeae* produces Protein II.

3. A. Antigenic recognition allows a highly specific immune response.

   B. By antigenic variation a host changes its immune response.

4. A. IgA antibodies destroy IgA proteases.

   B. *Neisseria gonorrhoeae* makes an IgA protease.

5. A. Serum resistance acts against a complement system.

   B. Coagulase is an enzyme that destroys complement proteins.

6. A. Siderophores add iron to iron-transport proteins.

   B. Transferrin is an iron-transport protein.

Six: Multiplying in the Host

A. Matching - Each choice is used once.

1. endotoxin		A.	component used in a diphtheria vaccine
2. hypersensitivity		B.	condition caused by *Neisseria*
3. interleukin-1		C.	cytokine causing fever
4. fibrin		D.	disrupts adenyl cyclase system
5. meningococcemia		E.	exaggerated immune response
6. Ptx		F.	forms blood clots
7. tetanus toxin		G.	found in all Gram-negative bacteria
8. hyaluronidase		H.	cements cells together

B. Complete each of the following with the correct term or terms.

1. _____ _____ is an enzyme that potentiates the effect of the pertussis toxin in eukaryotic cells.

2. Cytolysins attack cell _____.

3. Collagen is a major structural component of _____ tissue.

4. Coagulase stimulates the conversion of _____ to _____.

5. Streptokinase is used to dissolve _____ _____ .
6. Viruses destroy host cells from within by the process of _____ .
7. During a _____ infection, a virus is dormant for years in a host cell.
8. The rabies virus produces _____ in infected nerve cells.

Seven: Leaving the Body

A. Name the portal of exit for each of the following.
1. respiratory pathogens
2. gastrointestinal pathogens
3. sexually transmitted pathogens
4. pathogens from arthropod vectors

Emerging and Reemerging Infections Diseases

1. How can crossing a species barrier lead to devastating diseases in humans?

Multiple Choice: Review

1. A pathogen must have a reservoir in order to
   A. attach to a host.
   B. gain a good source of food.
   C. overcome the action of antibiotics.
   D. survive outside the human host.

2. Select the correct association.
   A. *Clostridium*/produces capsules
   B. fish/reservoir for Lyme disease
   C. *Salmonella*/causes chlolera
   D. *Streptococcus*/throat infection

3. *Bordetella pertussis* is spread between hosts by
   A. blood transfusions.
   B. contact between mucous membranes.
   C. contaminated food.
   D. respiratory droplets.

4. STD infections involve the contact of _____ membranes.
   A. cutaneous
   B. mucous
   C. serous
   D. synovial

5. A species of *Vibrio* is the causative agent of
   A. botulism.
   B. cholera.
   C. gonorrhea.
   D. malaria.

6. Biological vectors usually transmit infection by
   A. adhering to the mucous membranes.
   B. biting the host.
   C. entering through the gastrointestinal tract.
   D. entering through the respiratory tract.

7. FHA is a type of
   A. adhesin.
   B. antibody.
   C. antigen.
   D. white blood cell.

8. The species of *Plasmodium* enter the hosts'
   A. muscle cells.
   B. nerve cells.
   C. red blood cells.
   D. white blood cells.

9. Select the protein that does not transport iron in the human body.
   A. ferritin
   B. lactoferrin
   C. sideroferrin
   D. transferrin

10. Select the incorrect association.
    A. coagulase/breaks down fibrin
    B. hyaluronic acid/cements host cells together
    C. interleukin-1/stimulates fever production
    D. streptokinase/dissolves blood clots

Answers

Correlation Questions

1. The incubatory carriers are recognized by an absence of symptoms. The chronic carriers are detected by the presence of the causative microbe in their bodies, often harboring it for years.

2. Begin by establishing the complete geographical range of the animal reservoir in the wild. Then develop a strategy to eliminate the animal. For example, a deer population could be eradicated by increased hunting practices in its area.

3. The hypodermis, residing under the dermis of the skin, is highly vascular. The blood flow in the hypodermis is necessary to distribute the vaccine throughout the body.

4. This pathogen is depleting a major way that the body stores and circulates iron throughout the body. With less iron available to make functional red blood cells, the oxygen-carrying power of the body is diminished (anemia).

The Seven Capabilities of a Pathogen
One: Maintaining a Reservoir

A. 1. The disease-causing capabilities of a microorganism, enabling it to cause an infection.

2. A person who is a human reservoir for a pathogen. The person is infected but has not yet developed symptoms.

3. A person who is a human reservoir for a pathogen. The person harbors the pathogen for months or years but never becomes sick.

4. A human disease caused by a pathogen with an animal reservoir.

5. An animal that provides an environment for the survival of a pathogen.

6. A nonliving source that provides an environment for the survival of a pathogen.

Two: Getting To and Entering a Host

A. 1. True

2. False; The infectious dose is the number of cells needed to cause *infection* in 50% of the test animals.

3. False; The lethal dose is the number of cells needed to cause *death* in 50% of the test animals.

4. True

5. True

6. False; A fomite is an inanimate object serving as a vehicle for transmitting a pathogen.

7. True

8. False; STD transmission involves the contact of mucous membranes.

9. False; Herpes simplex I is caused by a virus.

10. False; A perinatal infection occurs during the passage through the birth canal or shortly after birth.

11. True

12. True

13. False; Airborne microbes, causing infections, may settle on dust particles and become airborne again.

14. True

15. False; A parenteral infection occurs by transmission through a break in the skin or mucous membranes. By these breaks, the infecting microbe enters the blood of the human host.

Three: Adhering to a Body Surface

A. 1. adhesins 2. trophism 3. cilia 4. filament 5. rectal, conjunctival

Four: Invading the Body

A. 1. *Bordetella pertussis* and *Streptococcus pneumoniae*
2. phagocytosis
3. They engulf and destroy invading microbes.
4. It can overcome phagocytosis by the host cells they invade. It survives and multiplies in the phagocyte.
5. Invasins are proteins on a pathogen's surface that stimulate phagocytosis in the cells they attack. Normally these cells are not phagocytic. The phagocytosis of the attacked cells helps the invading microbes to enter the host cells..
6. Intracellular pathogens stay inside host cells, multiplying there and not spreading.

Five: Evading the Body's Defenses

1. A 2. B 3. A 4. B 5. A 6. B

Six: Multiplying in the Host

A. 1. G 2. E 3. C 4. F 5. B 6. D 7. H 8. A

B. 1. adenyl cyclase 2. membranes 3. connective 4. fibrinogen to fibrin 5. blood clots 6. autolysis 7. latent 8. Negri bodies

Seven: Leaving the Body

A. 1. Respiratory pathogens leave through the nose by respiratory secretions.
2. Gastrointestinal pathogens leave by the anus.
3. STD pathogens leave across the mucous membrane surfaces of the genital tract.
4. Pathogens from arthropod vectors leave via a drop of blood.

Emerging and Reemerging Infections Diseases

1. Some pathogens, that previously infected nonhuman species, cross over and infect another species, namely humans.

Multiple Choice: Review

1. D 2. D 3. D 4. B 5. B 6. B 7. A 8. C 9. C 10. A

Chapter 16
The Immune System: Innate Immunity

Key Terms

innate immune system

adaptive immune system

nonspecific interior defenses

leukocytes

plasma

complement

phagocyte

alternative pathway

macrophage

neutrophil

monocyte

cell surface receptors

phagolysosome

cytokine

inflammation

mast cell

erythema

chemotactic

inflammatory repair

interferon

fibroblast

natural killer cell

Study Tips

1. A study of physiology will you help to understand the concepts in this chapter. Review the functions of macrophages and mast cells as well as hematology.

2. Can you reserve some time in your biology lab to study the human skeleton? Use the skeleton as a reference to study the various sites of blood cell production.

3. Also, use the lab as an opportunity to study the following cells under the microscope: mast cell, macrophage, neutrophil, eosinophil, basophil, lymphocyte, and monocyte.

4. Do you understand how mathematical concepts can pertain to the concepts in this chapter? Ask your instructor for some examples or make up your own study questions. For example, if the total white blood cell count in a patient is 8000 cells per cubic mm, what is the most frequent

white blood cell type in this total? If its number is 5000 of the total, how is its percentage expressed in a differential white blood cell count?

5. Fever is rarely harmful to the human body. However, is it an effective means of protecting the body? Form a discussion group with your classmates to discuss this topic.

Correlation Questions

1. What is the relationship between erythema and phagocytosis as protective mechanisms in the human body?

2. The skin is sometime called the first line of defense in the body. The innate immune system is a second line of defense. Explain.

3. Do you think that the different kinds of white blood cells are different because of variation in the genes they inherit or is there another explanation?

4. Aspirin blocks the effect of inflammatory mediators, such as prostaglandins, in the body. What beneficial effect does this have on body function?

The Body's Three Lines of Defense Against Infection

Completion
1. The human body has __(number)__ lines of defense against invading microorganisms.
2. The innate immune system acts to protect the body if invading microorganisms penetrate the three _____ defenses.
3. The _____ immune system is the third and last line of defense to protect the body.

The Innate Immune System

Completion
1. _____ are the white blood cells.
2. White blood cells in the circulation are suspended in the _____, the liquid part of the blood.
3. The first response of the innate immune system is the activation of _____.
4. The phagocytes are a class of _____, immune cells that can carry out phagocytosis.
5. Hydrolysis off a complement means the _____ with _____.
6. C3 convertase is a(n) _____ that rapidly forms molecules of C3b.
7. Macrophages and _____ are the two kinds of phagocytes.
8. _____ are the most abundant kind of white blood cell.
9. _____ is the first step of phagocytosis.
10. A white blood cell can engulf a bacterium by extending its _____.
11. A _____ is a membrane-bound package of lethal chemicals inside a phagocyte.

12. Macrophages release compounds called _____ which signal other components of the immune system to respond to invaders.
13. _____ cells can degranulate to release inflammatory mediators.
14. The _____ of capillaries increases their size and blood flow that they supply to an area.
15. Erythema can change the temperature of a body area by making it _____.
16. Chemotaxis is a chemical _____.
17. The first step in inflammatory _____ is clean up.
18. By acute inflammation human tissues can be _____.

Defense Against Viral Infections

Label each of the following statements as true or false. If false, correct it.
   1. There are three main classes of interferons.
   2. Beta interferon is produced by red blood cells.
   3. Cells that are infected by a virus are stimulated to produce interferons.
   4. Interferons are virus specific.
   5. NK cells are leukocytes.
   6. NK cells are effective at fighting bacterial cells but not viruses.

Multiple Choice: Review

1. Inflammatory mediators usually cause the
   A. vasoconstriction of blood vessels.
   B. vasodilation of blood vessels.

2. Complement is
   A. part of the innate immune system.
   B. not part of the innate immune system..

3. Histamine increases the diameter of blood vessels. Therefore, it is a
   A. vasoconstrictor.
   B. vasodilator.

4. During erythema, there is a(n)_____ bloodflow to the skin.
   A. decreased
   B. increased

5. Phagocytes destroy bacteria by
   A. crenation.
   B. phagocytosis.

6. Select the leukocyte that gives rise to macrophages.

    A. basophil
    B. eosinophil
    C. neutrophil
    D. monocyte

7. Select the most abundant white blood cell..

    A. basophil
    B. eosinophil
    C. neutrophil
    D. monocyte

8. Select the last step of phagocytosis.

    A. expulsion.
    B. chemotaxis
    C. digestion
    D. recognition

9. Activation of complement in the innate response proceeds through the _____ pathway.

    A. alternate
    B. classical

10. Interferons are

    A. virus-nonspecific.
    B. virus-specific.

Answers

Correlation Questions

1. The red area of erythema in body regions indicates increased blood flow to that area. This improved circulation brings more phagocytic cells to that area, fortifying the area with more immune cells to protect that area.

2. The skin is one of the first barriers an invading microbe encounters when it enters the body. However, if the skin is defeated, secondary immune lines can serve as backups to protect the body.

3. All cells of the body have the same genome, inherited and preserved by mitosis. The difference in the variety of human cells, including the different kinds of white blood cells, is due to differential gene activation. Some genes, if activated, will lead to the development of a neutrophil. A macrophage has these same genes. Its genome is identical to the neutrophil. However, activation of a different group of its genes causes it to take a different line of development.

4. Aspirin can block the stages of the inflammatory response. This can be desirable if this response is too powerful, possibly damaging the body tissues.

The Body's Three Lines of Defenses

1. three   2. surface   3. adaptive

The Innate Immune System

1. Leukocytes 2. plasma 3. complement 4. leukocyte 5. splitting, water 6. enzyme
7. neutrophils 8. neutrophil 9. Recognition 10. pseudopodia 11. lysosome 12. cytokines
13. mast 14. Dilation 15. warmer 16. attraction 17 repair 18. damage

Defense Against Viral Infections

1. True
2. False; It is produced by fibroblasts.
3. True
4. False; It is virus nonspecific.
5. True
6. False; They effectively fight viral infections.

Multiple Choice: Review

1. B   2. A   3. B   4. B   5. B   6. D   7. C   8. A   9. A   10. A

Chapter 17
The Immune System: Adaptive Immunity

The Adaptive Immune System: An Overview
    Lymphocytes
    Humoral Immunity
    Cell-Mediated Immunity
    Lymphoid Immunity
    Antigens
The Humoral Immune Response
    Generating Antibody Diversity
    Structure of Antibodies
    B-Cell Activation
    Reactivation: Immunological Memory
    Action of Antibodies
The Cell-Mediated Immune Response
    Antigen Recognition
    T-Cell Activation
    T-Cell Response
Natural Killer Cells
Immune Tolerance
Types of Adaptive Immunity
The Role Of The Adaptive Immune System in T. L.'s Recovery
Summary

Key Terms

differentiate	memory B cell
antigen	interleukin
antibody	immunological memory
humoral immunity	vaccine
cell-mediated immunity	antitoxin
bone marrow	membrane-attack complex
stem cell	dendritic cell
lymphatic circulation	perforin
immune tolerance	clonal depletion
epitope	apotosis
immunocompetent	autoimmune disease
immunoglobulin	active immunity
clonal selection	passive immunity
plasma cell	

Study Tips

1. Continue to answer the several kinds of questions at the end of the chapters. Discuss the answers with your classmates. Ask your instructor to post an answer key to check for the accuracy of your responses.

2. Is there a human torso model in your biology lab? Refer to it to locate the different organs in the body with lymphatic tissue.

3. If there is a long bone in your biology lab, study it to locate the sites of blood cell formation from stem cells in the marrow cavity.

4. Use a molecular model kit to build the basic makeup of an antibody molecule. If such a kit is unavailable, sketch the basic makeup from memory.

5. Continue your practice with molecular modeling with various activities posted on the Internet. Start with the search term *molecular models*.

Correlation Questions

1. An antibody, which is divalent, has a pentameter structure. How is this an advantage as it carries out its role in the body?

2. An undergraduate student is training for a career in medical technology. How does a study of the different kinds of immune cells contribute toward that training?

3. Are the precursor cells for immune cells the only kind of stem cells in the body? What are other kinds of stem cells that function during human development?

4. How can the administration of a vaccine be dangerous to a person's health?

The Adaptive Immune System: An Overview

A. Complete each of the following statements with the correct term.
   1. _____ are the most important kind of cell in the adaptive immune system.
   2. The cells of lymphoid tissue _____ as they mature and acquire special characteristics and functions.
   3. There are __(number)__ kinds of lymphocytes.
   4. _____ are foreign molecules that can enter the body through the attack of pathogens.
   5. T and B cells produce defensive proteins called _____.
   6. Another name for antibody-mediated immunity is _____ immunity.
   7. Antibodies protect the body against invading pathogens three ways. In addition to opsonization and activation by complement, they _____ toxins by binding to them.
   8. _____ cells confer cell-mediated immunity.

9. Bone _____ is a site in the body where stem cells produce B and T lymphocytes.

10. Most lymphocytes are stored in the _____ _____.

11. Lymph of the _____ duct is drained into the heart.

12. Almost all _____ molecules are strong antigens.

13. _____ and _____ are molecules that are devoid of antigenic activity.

14. The ability not to react to the body's own molecules is called _____ _____.

15. Lymphocytes recognize antigens by the antigen _____ on their cell surface.

16. Lymphocytes recognize epitopes by their distinctive _____.

The Humoral Immune Response

A. Label each of the following statements as true or false. If false, correct it.
1. During differentiation in the bone marrow, B cells become immunocompetent.
2. There are many more human genes than antibodies.
3. During differentiation, individual B lymphocytes make their own unique antibodies through genetic recombination.
4. An antigen must bind to the surface of an antibody in order to undergo activation.

B. Select the correct statement from each of the following pairs.
1. A. Antibody molecules lack a constant region.
   B. Antibody molecules have two heavy chains and two light chains.
2. A. Antibodies are Y-shaped molecules.
   B. The variable region of an antibody determines what it might do functionally.
3. A. There are five classes of antibodies.
   B. There are ten classes of antibodies.
4. A. Antibodies are glycoproteins.
   B. Antibodies are nucleic acids.

C. Match each class of antibody to its correct description.
1. IgA
2. IgD
3. IgE
4. IgG
5. IgM

A. It protects the mucous membranes of the body.
B. Its function is unknown.
C. It causes several kinds of cells to degranulate.
D. It acts as an opsonin.
E. It is active during the earliest phase of the primary immune response.

D. Label each of the following as true or false. If false, correct it.
1. Clonal selection produces a group of identical B cells.
2. Lymphokines are substances that kill lymphocytes.

3. Macrophages secrete interleukin-1.

4. B-cell activation is a one-step process.

5. When an infection is over, plasma cells continue to circulate in the body.

6. A secondary immune response takes weeks to develop.

7. A secondary immune response is initiated by memory B cells.

8. The fit between an antigen and its specific antibody is somewhat variable.

9. Antitoxins bind to toxins that act as antigens.

10. The classical complement pathway is active by the Fc region of an antibody.

The Cell-Mediated Immune Response

A. True/False: Four of the following eight statements are true. Select the correct statements.

1. T cells are able to control intracellular infections.

2. T cells cannot produce lymphokines.

3. Dendritic cells are antigen-presenting cells.

4. MHC are specialized genes making antibody molecules.

5. T lymphocytes are in a resting state until their antigenic receptors encounter and bind to a matching antigen fragment.

6. Human cells cannot signal that they have been infected by a virus.

7. Perforins secreted by TC cells mark infected cells for antibody activation.

8. Human cells can signal that they have been infected by bacteria.

B. Completion - Complete each of the following with the correct term.

1. Natural killer cells can function without recognizing a(n) _____.

2. Non-B non-T lymphocytes are sometimes called natural _____ cells.

3. NK cells target and kill human cells by _____ them.

4. NK cells can respond against _____ organs.

C. Completion - Complete the following paragraph with the correct terms.

Differentiating __1__ cells in the bone marrow and __2__ cells in the thymus have not fully specialized. They undergo __3__, which is programmed cell death. By reacting with self antigens, these cells are __4__ during development. This process is called clonal __5__.

D. Short Answer

1. Name the four basic kinds of adaptive immunity.

Multiple Choice: Review

1. Humoral immunity means that the antibodies are
    A. constantly changing.
    B. destroyed in the serum.
    C. dissolved in fluid.
    D. reacting with antigens

2. Select the primary lymphoid organs.
    A. bone marrow and thymus
    B. bone marrow and lymph nodes
    C. lymph nodes and spleen
    D. spleen and thymus

3. An epitope is a
    A. differentiated B cell.
    B. differentiated T cell.
    C. small part of an antibody.
    D. small part of an antigen.

4. Which cell becomes immunocompetent?
    A. T cell
    B. B cell

5. By which process do B cells proliferate?
    A. clonal selection
    B. immune tolerance

6. Which cell type can differentiate into a plasma cell?
    A. T cell
    B. B cell

7. Which immune response generates more antibodies?
    A. primary
    B. secondary

8. For an antibody molecule, the variable regions are at the
    A. end of an X shaped molecule.
    B. end of a Y shaped molecule.
    C. middle of an Y shaped molecule.
    D. middle of an X shaped molecule.

9. Which class of antibodies protect mucosal surfaces?
   A. IgA
   B. IgB
   C. IgE
   D. IgM

10. Which type of immunity is stimulated by vaccines?
    A. artificially acquired active
    B. artificially acquired passive
    C. naturally acquired active
    D. naturally acquired passive

Answers

Correlation Questions

1. This type of molecule has numerous binding sites. This maximizes its ability to bind with numerous antigen molecules, enhancing its power to inactivate them.

2. Each cell has a distinctive appearance. Based on differences in size, nucleus morphology, staining color, and presence/absence of cytoplasmic granules, each kind of immune cell is unique. A medical technologist must be able to distinguish a neutrophil (multilobed nucleus, blue cytoplasmic granules) from an eosinophil (bilobed nucleus , red cytoplasmic granules). Each immune cell has unique abilities and its changing concentration in a subject is meaningful, often being used in diagnosis.

3. A stem cell is not confined to the bone marrow cells from which most immune cells are produced. It can be any undifferentiated cell, often found early in fetal development before body cells have differentiated and taken on specific roles.

4. Normally a vaccine has an attenuated form of a microorganism. Its presence is sufficient to stimulate the immune system without causing its disease. However, rarely the microbe may be virulent to cause a disease.

The Adaptive Immune System: An Overview

1. lymphocyte 2. differentiate 3. three 4. Antigens 5. antibodies 6. humoral 7. neutralize 8. T 9. marrow 10. lymph node 11. thoracic 12. protein 13. lipid/nucleic acid 14. immune tolerance 15. receptors 16. shape

The Humoral Immune Response

A. 1. True

2. False; There are 30,000 human genes. There are 100,000 million human antibodies.
3. True
4. True

B. 1. B 2. A 3. A 4. A

C. 1. A 2. B 3. C 4. D 5. E

D. 1. True
   2. False; Lymphokines are molecules produced by lymphocytes and are necessary for the immune response.
   3. True
   4. False; B cell activation is an incremental process.
   5. False; Plasma cells die when an infection is over, but memory cells circulate.
   6. False; The secondary immune response requires only a few days.
   7. True
   8. True
   9. True
   10. True

The Cell-Mediated Immune Response

A. 1, 3, 5, 8
B. 1. antigen 2. killer 3. lysing 4. transplanted
C. 1. B 2. T 3. apotosis 4. eliminated 5. deletion
D. 1. naturally acquired active immunity, artificially acquired active immunity, naturally acquired passive immunity, artificially acquired passive immunity

Multiple Choice: Review

1. C 2. A 3. D 4. B 5. A 6. B 7. B 8. B 9. A 10. A

Chapter 18
Immunological Disorders

Immune System Malfunctions
Hypersensitivity
    Type I: Anaphylactic Hypersensitivity (Allergy)
    Type II: Cytotoxic Hypersensitivity
    Type III: Immune-Complex Hypersensitivity
    Type IV: Cell-Mediated (Delayed) Hypersensitivity
Organ Transplantation
    Immunosuppression
Immunodeficiencies
    Congenital Immunodeficiencies
    Acquired Immunodeficiencies
Cancer and The Immune System
Summary

Key Terms

hypersensitivity

allergen

allergic rhinitis

Grave's disease

erythrocyte

hemolytic disease

exchange transfusion

serum sickness

granulomatous reaction

isograft

xenograft

acquired immunodeficiency

lymphomas

tumor

immunodeficiency

anaphylaxis

asthma

Goodpasture's syndrome

transfusion reaction

Rh-immunized

systemic lupus erythematosus

contact hypersensitivity

autograft

allograft

congenital

leukemia

transformation

immune surveillance theory

Study Tips

1. Interview a local health professional on your campus or nearby community. This communication can provide you with additional insights to understand the concepts in this chapter.

2. Arrange to study transverse sections of blood vessels and bronchi/bronchioles in lab. Knowledge of their histology (tissue makeup) will help you to understand some of the reactions described in the chapters on immunity. For example, when the smooth muscle in the walls of these structures contracts, these structures constrict. Relaxation of the muscle results in dilation.

3. Use of an artificial blood typing kit in lab can help you to visualize the antigen-antibody reactions of the ABO and Rh blood groups.

4. Continue to refer to the human torso model in lab. Locate structures such as the thymus gland, spleen, and lymph nodes.

5. Consult the Internet to add to knowledge about immunological disorders. Use the key terms as search terms.

Correlation Questions

1. How can advances in endocrinology contribute to treating Grave's disease. Why is there less potential from this field to treat Goodpasture's syndrome?

2. Blood type O is often called the universal donor. How is this description inaccurate?

3. How is the ABO blood typing system more complex than the Rh typing system?

4. How can inhibiting angiogenesis, the development of blood vessels serving tissues, be a safer treatment for cancer compared to the use of chemotherapy?

Immune System Malfunctions/Hypersensitivity

A. Select the correct statements about anaphylactic hypersensitivity.
1. Its more common name is allergy.
2. Sensitization by an allergen leads to a harmful allergic reaction with future exposure to an allergen.
3. The attachment of IgA antibodies to basophils is a major part of this response.
4. Histamine, an inflammatory mediator, leads to massive vasoconstriction of blood vessels.
5. Compared to histamine, leukotrines act slowly.
6. Inflammatory mediators produce the signs and symptoms.
7. The cross section of the larynx increases during this response.
8. Edema of the intestinal tract develops.
9. Fluid is lost rapidly from the circulation.
10. As an effective treatment, the administration of epinephrine constricts blood vessels and relaxes bronchial muscle.
11. This type of reaction is very rare.
12. Inhaled allergens can cause allergic rhinitis.
13. Hay fever is primarily mediated by histamine.
14. Inhaled allergens that affect the upper respiratory tract cause asthma.
15. Bronchi narrow during the response in asthma.
16. Histamine is mainly responsible for asthma.
17. Respiratory allergies are detected by skin testing.
18. Desensitization is effective at inhibiting respiratory allergens.

19. Ingested allergens are relatively harmless.
20. Penicillin is a hapten that can become an allergen.

B. Select the correct statements about cytotoxic hypersensitivity.
   1. IgM and IgG antibodies bind abnormally to body cells.
   2. This response may activate the terminal complement pathway.
   3. All autoimmune diseases are caused by type II sensitivity.
   4. Grave's disease develops when antibodies bind to cells that produce human growth hormone.
   5. In Goodpasture's syndrome, body cells are killed.
   6. The ABO blood groups are name by the antibodies present.
   7. The ABO antibodies are of the type IgM class.
   8. Agglutination is the lysing of erythrocytes.
   9. A person with type B blood can receive type O blood without agglutination.
   10. A person with type A blood can donate that blood to a person with type AB blood without agglutination.
   11. Rh negative blood lacks the Rh factor.
   12. Production of anti-Rh antibodies requires exposure to Rh-positive blood.
   13. The hemolytic disease of the newborn involves an Rh negative baby.
   14. The hemolytic disease of the newborn involves an Rh positive mother.
   15. The hemolytic disease of the newborn can be prevented by preventing the Rh-immunization of the Rh negative mother.

C. Select the correct statements about Immune-complex hypersensitivity.
   1. IgG and IgM antibodies react with a person's body cells.
   2. Circulating, soluble antigen-antibody complexes initiate this kind of reaction.
   3. Complement is activated in capillaries.
   4. SLE can be effectively treated.
   5. Rheumatoid arthritis is an autoimmune disease.

D. Select the correct statements about cell-mediated hypersensitivity.
   1. It is initiated by certain TH lymphocytes.
   2. This response occurs within a few minutes after exposure.
   3. The granulomatous reaction causes tissue damage.
   4. Poison ivy occurs by this type of hypersensitivity.
   5. Poison oak is caused by this type of hypersensitivity.
   6. TD cells release lymphokines leading to granuloma formation.

Organ Transplantation

A. Matching - Each transplant description is used once.

1.	allograft	A.	transplant between two species
2.	autograft	B.	transplant from one identical twin to another
3.	isograft	C.	transplant from one region of the body to another.
4.	xenograft	D.	transplant between members of the same species

Immunodeficiencies

A. Complete each of the following statements with the correct term or terms
1. SCID involves disorders that disable both _____ cell and _____ cell immunity.
2. Patients with X-linked agammaglobulinemia have a severe deficiency of _____ cells.
3. DiGeorge's syndrome affects only _____ cell function.
4. Lymphoma is cancer of the _____ _____.

Cancer and the Immune System

A. Short Answer
1. How many cytotoxic T cells, NK cells, and macrophages play a role in controlling cancer?
2. What is the immune surveillance theory?

Multiple Choice: Review

1. Hypersensitivity is a _____ response.
    A. rapid
    B. slow

2. During an allergic reaction, bronchi
    A. constrict
    B. dilate

3. The action of inflammatory mediators leads to
    A. blood fluid gain.
    B. blood fluid loss.

4. During asthma inhaled allergens affect the _____ respiratory tract.
    A. upper
    B. lower

5. Grave's disease involves the malfunction of the _____ gland.

   A. adrenal
   B. pituitary
   C. thymus
   D. thyroid

6. Bloodtype A can donate to bloodtypes _____ without causing agglutination.

   A. A and O
   B. A and AB
   C. B and O
   D. B and AB

7. A person who is Rh positive usually __1__ the Rh factor and __2__ the anti-Rh antibody.

   A. 1 - possesses, 2 - possesses
   B. 1 - possesses, 2 - lacks
   C. 1 - lacks, 2 - possesses
   D. 1 - lacks, 2 - lacks

8. Rheumatoid arthritis is a type _____ hypersensitivity.

   A. I
   B. II
   C. III
   D. IV

9. Which type of organ transplant only involves one organism's body?

   A. allograft
   B. autograft
   C. isograft
   D. xenograft

10. The term *congenital* means

    A. allergic.
    B. at birth.
    C. disease-producing.
    D. rheumatoid.

Answers

Correlation Questions

1. Grave's disease develops from hyperactivity of the thyroid gland, part of the endocrine system. Goodpastures's syndrome does not have a well-known endocrinological basis.

2. There are dozens of antigen-antibody systems that type the blood. Type O is the universal donor only among the ABO blood types. To avoid transfusing an incompatible blood type, other blood typing systems must also be checked.

3. The ABO system is based on two antigens, with four possible combinations. The Rh system has one antigen that is either present (Rh positive) or absent (Rh negative).

4. Cancer cells can be starved if their blood supply is removed. Potentially this does not have the side effects of chemotherapy, which slows the rate of cancer cell division through chemical treatment.

Immune System Malfunctions/Hypersensitivity

A. 1, 2, 5, 6, 8, 9, 10, 12, 13, 15, 17, 20
B. 1, 2, 5, 7, 9, 10, 11, 12, 15
C. 2, 3, 5
D. 1, 3, 4, 5, 6

Organ Transplantation

A. 1. D 2. C 3. B 4. A

Immunodeficiencies

A. 1. B and T
2. B
3. T
4. lymph nodes

Cancer and the Immune System

1. They destroy antigen-marked cells that are transformed in the development of cancer.

2. The transformation of cancer cells occurs often, but the immune system usually eliminates these cancer cell.

Multiple Choice: Review

1. A 2. A 3. B 4. B 5. D 6. B 7. B 8. C 9. B 10. B

Chapter 19
Diagnostic Immunology

Diagnostic Immunology

Detecting Antigen-Antibody Reactions
    Precipitation Reactions
    Agglutination Reactions
    Complement Fixation Reactions
Immunoassays
Fluorescent Antibodies
Summary

Key Terms

diagnostic immunology

serology

test reagent

equivalence zone

double diffusion method

electrophoresis

direct agglutination reaction

titer

indicator system

immunofluorescence assay

enzyme-linked immunosorbent assay

fluorescein isothiocyanate

monoclonal antibody

serum

precipitation reactions

immunodiffusion tests

radial diffusion

Western blot

indirect agglutination reaction

complement fixation

radioimmunoassay

solid phase immunosorbent assay

fluorescent antibody

Study Tips

1. Visit a local faculty member or research scientist conducting studies in the fields described in this chapter. If you cannot locate this person, ask your instructor for advice in your search.

2. Are any of your lab activities offering opportunities to study the concepts of diagnostic immunology? Ask your instructor for examples.

3. Use molecular models to demonstrate the various bonding patterns between antigens and antibodies.

4. The Internet provides exercises to make molecular models. Use this resource to learn more about the chemical structure of antigens and antibodies.

Correlation Questions

1. Connective tissues in animals include compact bone, cartilage, and blood. In each, cells are arranged in an intercellular matrix. In bone, for example, the matrix contains salts of calcium and phosphorus. How is blood, with its makeup cellularly and intercellularly, a unique kind of connective tissue?

2. Why are agglutination tests performed on a slide rather than in a test tube?

3. Why do antigen and antibody molecules need more than one binding site if a precipitation reaction is to occur?

4. When is a quantitative test needed instead of a qualitative test?

Diagnostic Immunology

A. Completion - Complete each of the following with the correct term.
1. Diagnostic immunology is a field that uses antigen -antibody reactions to diagnose _____.
2. The pregnancy test uses _____ antibodies.
3. A blood sample is about 45 percent cellular, by volume. The percent that is serum is about _____ percent.
4. In a diagnostic immunology test, _____ in a test reagent are used to detect antigens.
5. A test is conducted to learn if a person is infected by hepatitis A. Diagnosis involves detecting antibodies that a person makes against the _____ molecules of a virus.

Detecting Antigen-Antibody Reactions

A. Select the correct statements among the following.
1. All serologic tests involve a reaction between an antigen and an antibody.
2. After a chemical reaction, the contents in a test tube changes from cloudy to clear. This is an indication of an antigen-antibody reaction.
3. A certain antigen and antibody each has one binding site. This will allow them to react and build a molecular lattice.
4. For an immunodiffusion test, an antibody in a well spreads through the surrounding gel. Its concentration in the gel increases with increasing distance from the well.
5. In a simple immunodiffusion test, a precipitate between diffusing antigen and antibody will form in the zone of equivalence.
6. In a double diffusion test, two antigens form a continuous line on two sides of an antibody well. This result means that the antigens are identical.
7. In a radial diffusion test, an increasing ring in the size of a precipitate means increasing antigen concentration.
8. Immunoelectrophoresis is used to separate antigens in a mixture.

9. A western blot test is less sensitive that immunoelectrophoresis.

10. In the agglutination reaction to detect human blood type, antigen A reacts with antibody A and forms a clump of cells.

11. In the agglutination reaction to detect human blood type, antigen B is on the surface of human erythrocytes. It is detected by reacting with antibody A.

12. The Coombs test is a sensitive test.

13. Agglutination occurs when cells do not clump together.

14. A titer is the lowest dilution of a test serum that causes agglutination.

15. In a complement fixation test assays are qualitative.

Immunoassays

A. Label each of the following statements about immunoassays as true or false. If false, correct it.
1. They are extremely sensitive.
2. Antigens cannot be detected.
3. Antibodies can be detected.
4. To be performed successfully, an antigen-antibody product does not need to be separated from a tagged reactant.
5. ELISA are becoming the dominant immunoassays.

Fluorescent Antibodies

A. Short Answer
1. State a major use in the biology lab for fluorescent antibodies.

Multiple Choice: Review

1. In diagnostic immunology
   A. only an antigen can be used to detect an antibody.
   B. only an antibody can be used to detect an antigen.
   C. antigens and antibodies can be used to detect each other.
   D. antigens and antibodies are not used.

2. Serologic tests are
   A. quantitative only.
   B. qualitative only.
   C. quantitative and qualitative.
   D. neither quantitative nor qualitative.

3. For a precipitation reaction between an antigen and antibody, each molecule must have at least _____ binding sites.
   A. two
   B. four
   C. six
   D. eight

4. For an antigen-antibody reaction, cloudiness in a test tube indicates
   A. a precipitate and an antigen-antibody reaction.
   B. a precipitate without an antigen-antibody reaction.
   C. no precipitate and an antigen-antibody reaction.
   D. no precipitate without an antigen-antibody reaction.

5. In an electrophoretic field, the anode is the negative field. A positive molecule will migrate
   A. away from the anode.
   B. toward the anode.

6. Hemagglutination is a(n) _____ agglutination reaction.
   A. direct
   B. indirect

7. The ABO testing for human blood type is a _____ test.
   A. qualitative
   B. quantitative

8. A person's blood does not produce any agglutination in the ABO blood typing test. The person's ABO blood type is
   A. A.
   B. B.
   C. O.
   D. AB.

9. The complement fixation test is still very important for detecting a particular respiratory, syncytial
   A. alga.
   B. bacterium.
   C. protozoan.
   D. virus.

10. ELISA are _____ sensitive than most immunoassays.
    A. less
    B. more

Answers

Correlation Questions

1. In blood, the matrix is a unique liquid, namely the plasma. It has a complex chemical composition. Also, a tissue is defined as a group of similar cells working together. Not all blood cells are similar: erythrocytes, numerous kinds of leukocytes, and thrombocytes.

2. The surface provided by a slide is enough to view agglutination.

3. Two or more binding sites are needed in these molecules to allow bonding between them and the formation of a molecular lattice leading to precipitate formation.

4. Quantitative tests are needed for evaluating the severity or trend of a disease. For example, a changing concentration of eosinophils may indicated a parasitic infection. A changing level of IgE can indicate the development of an allergy. A qualitative test is needed to reveal the presence or absence of something: antigen A for blood type A or the presence of a species of *Neisseria* to indicate gonorrhea.

Diagnostic Immunology

A. 1. diseases 2. monoclonal 3. fifty-five 4. antibodies 5. antigen

Detecting Antigen-Antibody Reactions

A. 1, 5, 6, 7, 8, 10, 12

Immunoassays

A. 1. True
   2. False; Antigens and antibodies can be detected.
   3. True
   4. False; They antigen and antibody must be separated.
   5. True

Fluorescent Antibodies

A. 1. They are used to visualize specific antigens in tissues or on the surface of microorganisms.

Multiple Choice: Review

1. C 2. C 3. A 4. A 5. B 6. A 7. A 8. C 9. D 10. B

Chapter 20
Preventing Disease

Key Terms

epidemic

endemic

notifiable disease

prevalence rate

endoscopy

laparoscopy

prophylaxis

attenuated vaccine

toxoid

subunit vaccine

pandemic

sporadic

incidence rate

common source epidemic

bronchoscopy

urinary catheterization

immunization

inactivated vaccine

antitoxin

conjugate vaccine

Study Tips

1. Good students learn to anticipate questions asked by their instructor in a course. As you continue to study microbiology, this is a skill that you can develop. Try writing some questions on the content of this chapter.

2. Encourage your instructor to invite a representative from the local health department to address your class. If this representative cannot visit your class, try to meet this individual and conduct a personal interview. It will offer you additional insights about the topic of preventing diseases.

3. Continue to answer the several kinds of questions at the end of each textbook chapter.

4. Visit the Internet to learn more about currently-developed DNA-based vaccines. Start with the Web site quoted in the section of the text on DNA-based vaccines. Name some human diseases that might be prevented by the newest technological advances in this field.

Correlation Questions

1. Prevalence rates can be low for a disease that has a short duration in a population. Explain.
2. How is ecology, another branch of biology, relevant to the topic of disease prevention in microbiology?
3. Explain the advantage of active immunization compared to passive immunization when protecting a human.
4. Why are some diseases that specifically affect humans absent in other animal species?

Preventing Disease/Epidemiology

A. Complete each of the following statements with the correct term.

1. ________ is the name for a disease outbreak that affects many members of a population within a short time.
2. A ____________ is an epidemic that spreads worldwide.
3. A disease is __________ if it is always present in a population at about the same level.
4. A ___________ disease occurs only occasionally in a population.

B. Label each of the following statements as true or false. If false, correct it.

1. Vital statistics are kept by almost all governments.
2. Epidemiologists usually express health information as a rate.
3. The prevalence rate is the rate of acquiring a disease during a certain period of time.
4. The incidence rate is the rate of having a disease at any particular time.
5. If a disease currently infects 5 million people in the U.S., the incidence rate for that disease is one percent.

C. Match each of the following examples to the type of epidemiological study.

1. It tracks epidemic diseases.	A. descriptive
2. It studies nosocomial infections.	B. surveillance
3. It provides general information about diseases.	C. field
4. Koch conducted this type of study when seeking the cause of tuberculosis.	D. hospital
5. This investigates disease outbreaks.	
6. This type of study revealed that the HIV retrovirus caused AIDS.	
7. This study involves an infection introduced by laparoscopy.	
8. This study involves an infection resulting from a urinary catheterization.	

Public Health

A. Short Answer

1. One method of disease prophylaxis is limiting people's exposure to pathogens. What is the other method?
2. How can a physician learn if a recent outbreak of rabies has occurred in a local community?
3. What is the national public health agency of the United States?
4. How is the disease typhoid fever spread in a human community?
5. Name the two general kinds of microorganisms that cause diarrhea.
6. What kinds of diseases were transmitted through milk consumption before the use of pasteurization?
7. In food preparation, how can trichinosis be prevented?
8. What is the single most important personal habit that can prevent the spread of disease?
9. Currently, how effective is DDT at controlling insect populations?
10. Are sexually transmitted diseases a problem only in developing countries?
11. What is the current, main preventive measure against the spread of AIDS?
12. How is the microorganism causing diphtheria transmitted?

Immunization

A. Short Answer

1. List the several characteristics that make a vaccine safe and effective.
2. List the several routes by which vaccines can be administered.
3. What is an acellular vaccine?
4. What is an attenuated vaccine?
5. What are the drawbacks to inactivated vaccines?
6. What were the first acellular vaccines?
7. What are subunit vaccines?
8. What is the current state of research of DNA vaccines?
9. What is passive immunization?
10. What is the drawback of passive immunization?

Multiple Choice: Review

1. Which kind of disease occurs only occasionally in a population?
    A. endemic
    B. epidemic
    C. pandemic
    D. sporadic

2. Pellagra develops in humans due to a
   A. bacterial infection.
   B. mineral deficiency.
   C. vitamin deficiency.
   D. viral infection.

3. A country has a human population of 150 million. Three million people in this population acquire a disease over a specific time frame. The incidence rate for this disease is _____ percent.
   A. one
   B. two
   C. five
   D. six

4. Which type of epidemiological study provides general information about a disease?
   A. descriptive
   B. field
   C. hospital
   D. surveillance

5. Which type of epidemiological study investigates common source epidemics?
   A. descriptive
   B. field
   C. hospital
   D. surveillance

6. By urinary catheterization a device is placed in the
   A. bladder.
   B. kidney.
   C. ureter.
   D. urethra.

7. The term "prophylaxis" means
   A. immunization.
   B. infection.
   C. prevention.
   D. scrubbing.

8. Select the disease that is spread by an insect vector.
   A. cholera
   B. polio
   C. tetanus
   D. yellow fever

9. The development of Jenner's smallpox vaccine was an example of
   A. active immunization.
   B. passive immunization.

10. High titers of antibodies are administered to a person. This is an example of
    A. active immunization.
    B. passive immunization.

Answers

Correlation Questions

1. The prevalence rate is the rate of having a disease at any particular time. If the time span of the disease is short, the probability (calculated as a percentage) for the disease occurring is low.

2. Field and surveillance epidemiology depend on the investigation of organisms and their relationship to the environment—subjects of ecology.

3. Active immunization confers the potential for long-lasting protection and is not as short-lived as passive immunization.

4. Different species of animals have different genetic blueprints, possibly conferring immune mechanisms to some animals species and not others (e.g., humans).

Preventing Disease/Epidemiology

A. 1. epidemic 2. pandemic 3. endemic 4. sporadic

B. 1. True
   2. True
   3. False; This defines the incidence rate.
   4. False; This defines the prevalence rate.
   5. False; The incidence rate is 5 million divided by 250 million or 2.0 percent

C. 1. B 2. D 3. A 4. A 5. C 6. C 7. D 8. D

Public Health

A. 1. immunization
   2. The physician can contact the local health department.
   3. United States Public Health Service
   4. It is spread by human sewage.

5. The microorganisms are the bacterium and virus.

6. Examples included tuberculosis, typhoid fever, and scarlet fever.

7. It is prevented by cooking food adequately.

8. Hand washing is the most important preventive measure.

9. It is not currently effective.

10. They are a problem in both developing and developed countries.

11. Limited sexual exposure and the use of condoms are the preventive measure

12. It is spread by respiratory droplets.

Immunization

A. 1. It must be free of side effects; it must be highly immunogenic; it must be administered appropriately.

2. The routes include orally, subcutaneously, and intramuscularly.

3. They contain only certain components of a microorganism that are needed to stimulate a maximum immune response.

4. It is a genetically altered form of the microorganism. These live cells have limited ability to infect and cause illness.

5. It has dead microorganisms that cannot multiply in the host, making them less affected than attenuated vaccines.

6. They were toxoids.

7. They are purified protein components, which are antigenic and confer immunity.

8. It is an active field of research that will provide another means of immunization in the future. They are now being evaluated in human trials.

9. By passive immunity the antibody is introduced into the subject. The subject's immune system is not stimulated to produce the antibody. It is an antibody transfusion.

10. The effectiveness of passive immunity is short-lived.

Multiple Choice: Review

1. D 2. C 3. B 4. A 5. B 6. D 7. C 8. D 9. A 10. B

Chapter 21
Pharmacology

Principles of Pharmacology
    Drug Administration
    Drug Distribution
    Eliminating Drugs From The Body
    Side Effects and Allergies
    Drug Resistance
    Drug Dosage
Targets of Antimicrobial Drugs
    The Cell Wall
    Cell Membranes
    Protein Synthesis
    Nucleic Acids
    Folic Acid Synthesis
Antimicrobial Drugs
    Antibacterial Drugs
    Antimycobacterial Drugs
    Antifungal Drugs
    Antiparasitic Drugs
    Antimalarial Drugs
    Antiviral Drugs
    Interferons
Summary

Key Terms

antibiotic	drug
chemotherapeutic agent	antimicrobial agent
external therapy	systemic therapy
intravenous administration	intramuscular administration
oral administration	side effect
narrow-spectrum	broad-spectrum
penicillin-binding protein	combined therapy
disc-diffusion method	broth-dilution method
bacteriostatic	antimetabolite

Study Tips

1. Much of the information in this chapter deals with drugs and their physiological effects on microorganisms. Organize this information in the following table:

                Drug   Effect

2. Try to relate the information in this chapter to other topics you have studied in biology. For example, how are the different types of drug administration related to the anatomy of the human body? How do different drugs inactivate different parts of the anatomy and metabolism of microorganisms?

3. Continue to answer the several kinds of questions at the end of the chapter. Ask your instructor to post an answer key for these questions after you have tried them.

4. Start an Internet search in the field of pharmacology.

Correlation Questions

1. A drug consisting of protein molecules will not be effective if administered orally. Explain.

2. What do you think is the fastest means of introducing a drug into the body?

3. When is a broad-spectrum antibiotic a better choice for treatment compared to a narrow-spectrum antibiotic?

4. The extracellular fluid (blood plasma plus intercellular fluid) is often described as the internal environment of the body. Explain.

Principles of Pharmacology

A. Label each of the following statements as true or false. If false, correct it.
1. Pharmacology is the study of drugs.
2. Antimicrobial agents are used to treat infectious diseases.
3. Most infections require external therapy.
4. The oral administration of a drug usually has rapid, efficient effects.
5. Most drugs enter cells by passing through the phospholipid part of the cell membrane.
6. The binding of drugs to proteins in the blood plasma usually stops their effective administration.
7. Drug metabolism occurs mainly in the liver.
8. Penicillin is excreted rapidly from the body through the kidneys.
9. Most antimicrobial drugs have side effects.
10. Penicillin is effective at treating viral infections.
11. Narrow-spectrum antibacterial drugs affect only a single microbial group.
12. Ampicillin is a narrow-spectrum antibiotic.
13. Broad-spectrum antibiotics can significantly alter the normal biota of the body.
14. Acquiring an enzyme that destroys a drug is the least common mechanism by which microbes become resistant to drugs.
15. Many Gram-negative bacteria can pump tetracycline out of their cells.
16. An R factor encodes resistance to one specific antibiotic.

17. Generally in the environment, there is a rising tide of resistance to antibiotics by microorganisms.
18. By the Kirby-Bauer method, a clear halo develops around a disc impregnated with an antibiotic. The bacterium being tested is resistant to this antibiotic.
19. Disc-diffusion tests are relatively easy to perform.
20. The MIC shows the highest drug concentration that inhibits the growth of a bacterial population.

Targets of Antimicrobial Drugs

A. Match each drug to its principal target.
   1. flucytosine
   2. penicillin
   3. polymixin B
   4. sulfoanamides
   5. tetracycline

   A. cell membranes of fungi
   B. cell wall
   C. folic acid synthesis
   D. nucleic acids
   E. ribosome

Antimicrobial Drug

A. Complete each of the following with the correct term.
   1. The various kinds of penicillins differ by the identity of a chemical group in their structural formula called the _____ group.
   2. _____ _____ is the enzyme produced by some penicillin-resistant-pathogens.
   3. The cephalosporins are very effective against Gram - _____ bacteria.
   4. Among the families of antibiotics, the cephalosporins are closely related to the family of _____.
   5. The sulfonamides interfere with the ability of bacteria to synthesize _____ _____.
   6. The aminoglycosides are effective against Gram - _____ bacteria.
   7. Chloramphenicol can cause _____ anemia.
   8. The tetracyclines interfere with the ability of the ribosome to accept the molecule _____ during protein synthesis.
   9. Erythromycin is only effective against Gram - _____ bacteria.
   10. The quinolones block DNA _____ in bacteria.
   11. Vancomycin blocks _____ synthesis in bacteria.
   12. In comparing cellular regions, *Mycobacterium tuberculosis* is an _____ pathogen.
   13. Isoniazid is activated by the mycobacterial enzyme _____.
   14. Rifampin is made by _____ *mediterranei*.
   15. Nystatin is a member of the _____ family of antibiotics.
   16. Amphotericin B disrupts the cell membranes of _____.
   17. Griseofulvin is isolated from a fungus of the genus _____.

18. Metronidazole is effective against obligate _____ bacteria.
19. Amantadine is effective against the replication of _____.
20. Acyclovir is effective against DNA viruses of the _____ family.
21. Most reverse _____ inhibitors are nucleoside analogues.
22. _____ are small glycoproteins that stimulate other cells to make antiviral agents.

Multiple Choice: Review

1. A drug is introduced into the body by injection into the triceps brachii. This drug administration is
   A. EV.
   B. IV.
   C. IM.
   D. PO.

2. Most drugs are metabolized mainly by the
   A. liver.
   B. pancreas.
   C. small intestine.
   D. spleen.

3. Ampicillin is a _____ antibiotic.
   A. broad-spectrum
   B. narrow-spectrum

4. *Streptococcus pneumoniae* is mainly a _____ pathogen.
   A. gastrointestinal
   B. nervous system
   C. respiratory
   D. urinary tract

5. By the disc-diffusion method, the more sensitive a bacterium is to an antibiotic, the _____ the halo will be around the antibiotic disc.
   A. larger
   B. smaller

6. Streptomycin targets the _____ of bacterial cells.
   A. cell membrane
   B. cell wall
   C. nucleic acids
   D. ribosomes

7. Which antibiotics were the first to cure microbial infections after salvarsan?

    A. cephalosporins
    B. penicillins
    C. sulfoamides
    D. tetracyclines

8. Which antibiotic works against bacterial protein synthesis?

    A. aminoglycoside
    B. chloramphenicol
    C. penicillin
    D. tetracycline

9. Rifampin acts against

    A. Gram-positive bacteria only.
    B. Gram-negative bacteria only.
    C. Gram-negative and Gram-positive bacteria.
    D. viruses.

10. Select the antifungal agent that is a polyene.

    A. amphotericin B
    B. griesofulvin
    C. nystatin
    D. triazole

Answers

Correlation Questions

1. Enzymes, called proteases, in the digestive tract can degrade an ingested protein, rendering it useless before it can be absorbed into the blood in the small intestine.

2. IV injection introduces a drug into the circulation immediately to reach body cells. The drug does not need to clear the barriers by other means of administration.

3. If the causative agent of a disease is not known, a broad-spectrum offers more alternatives to defeat several possible microbes that could be infecting the body.

4. Although outside the cells (extracellular) this environment is found inside the organism. It bathes our cells internally and is the means by which drugs reach our cells.

Principles of Pharmacology

A. 1. True
   2. True

3. False; Most infections require systemic therapy.

4. False; Oral administration is slow and inefficient.

5. False; Most drugs are not lipid soluble.

6. False; Protein binding slows drug administration.

7. True

8. True

9. True

10. False; Viruses lack a cell wall and are resistant to penicillin.

11. True

12. False; Ampicillin affects Gram-negative and Gram-positive bacteria.

13. True

14. False; It is the most common mechanism.

15. True

16. False; It can encode resistance to several antibiotics.

17. True

18. False; This antibiotic stops microbial growth, as evidenced by the halo.

19. True

20. False; It shows the lowest drug concentration that inhibits bacterial growth.

Targets of Antimicrobial Drugs

A. 1. D 2. B 3. A 4. C 5. E

Antimicrobial Drugs

A. 1. *R*
    2. Beta-lactamase
    3. positive
    4. penicillins
    5. folic acid
    6. negative
    7. aplastic
    8. aminoacyl-tRNA
    9. positive
    10. replication
    11. peptidoglycan
    12. intracellular
    13. peroxidase

14. *Streptomyces*
15. polyene
16. fungi
17. *Penicillium*
18. anaerobic
19. viruses
20. herpes
21. transcriptase
22. interferons

Multiple Choice: Review

1. C 2. A 3. B 4. C 5. A 6. D 7. C 8. D 9. C 10. C

Chapter 22
Infections of the Respiratory System

Clinical Science
    Diagnosis
    Prognosis and Treatment
    Types of Infections
    Progress and Infection
Organization of Part IV
The Respiratory System
    Structure and Function
    Defenses and Normal Biota: A Brief Review
    Clinical Syndromes
Upper Respiratory Infections
    Bacterial Infections
    Viral Infections
Lower Respiratory Infections
    Bacterial Infections
    Viral Infections
    Fungal Infections
Summary

Key Terms

clinical presentation	symptoms
diagnosis	etiology
nasal cavity	adenoids
pharynx	trachea
larynx	pleura
epiglottis	rhinitis
pneumonia	rheumatic fever
diphtheria	rhinorrhea
quelling reaction	consolidated
psittacosis	filamentous hemagglutinin
tubercle	tuberculosis
syncytia	histoplasmosis

Study Tips

1. Most of the information in this chapter relates respiratory diseases to causative microorganisms. Organize this information by composing the following table:

               Disease   Microorganism

2. In addition to the information presented at the beginning of the chapter, study the human anatomy of the respiratory system in more depth. It will help you to understand the microorganisms and diseases discussed for the balance of the chapter.

3. An important emphasis of your study is to relate microbial activity to the different lines of defense that protect the respiratory system. This approach will help you to understand the various mechanisms of disease. For example, a microorganism can infect the body by overcoming the mucociliary system of the respiratory tract.

4. Continue to answer the several kinds of questions at the end of the chapter.

Correlation Questions

1. Why are smokers more susceptible to pneumococcal pneumonia?

2. Although lymphoid tissue, the tonsils must be surgically removed in some people. Why?

3. Infections of the upper respiratory tract can spread to the middle ear. How?

4. A person suffers from laryngitis. He is treated for a sore throat. Is this the correct mode of treatment?

Clinical Science

1. _____ is the cause of a patient's illness.
2. _____ is what a clinician can hear during a physical examination.
3. The combination of signs and symptoms in a patient are called clinical syndromes. The _____ implicates one particular structure.
4. Pharyngitis is a sore _____.
5. Acute, chronic, and _____ are terms that indicate the suddenness with which symptoms appear and how long they last.

The Respiratory System

A. Complete each of the following statements with the correct term describing the respiratory tract.
1. After the trachea, exhaled air enters the _____ next.
2. The adenoids are located in the _____.
3. From the primary bronchus, exhaled air enters the _____ next.
4. Oxygen and carbon dioxide diffuse across the walls of the _____.
5. The _____ and _____ lack the mucociliary system.

B. Short Answer
1. Describe the conditions of the upper respiratory tract that make it favorable for colonization of microbes.

2. What is rhinitis?
3. What is otitis media?
4. What is tachypnea?
5. How are the alveoli affected by pneumonia?

Upper Respiratory Infections

A. The following statements are about bacterial infections. Label each of the following statements as true or false. If false, correct it.
1. The epiglottis is prone to infection by *Haemophilus influenzae*.
2. Strains of *Haemophilus influenzae* can invade only human tissues.
3. Ampicillin is effective at treating respiratory infections caused by *Haemophilus influenzae*.
4. Alpha-hemolytic strains of *Streptococcus pyrogenes* can lyse red blood cells.
5. Humans are the only natural reservoir for group A streptococci.
6. Streptococcal pharyngitis usually lasts several weeks in an infected person.
7. Type B is the most virulent toxin causing streptococcal pharyngitis.
8. Rheumatic fever causes inflammation of many organs of the body.
9. Glomerulonephritis is an upper respiratory tract infection.
10. *Corynebacterium diphtheriae* is Gram-negative and rod-shaped.
11. Diphtheria can potentially cause lethal pharyngitis.
12. Erythromycin will neutralize the diphtheria toxin already formed in the body.
13. In the development of membranous pharyngitis, the lymph nodes of the neck swell.
14. Toxin-producing strains of *C. diphtheriae* are now rare.

B. The following statements are about viral infections. Label each statement as true or false. If false, correct it.
1. The common cold is caused by a virus.
2. Rhinoviruses are DNA viruses.
3. The only reservoir for rhinoviruses is human beings.
4. By IgA antibody production, people infected by a particular rhinovirus are protected against it for two months or less after the infection.
5. Antibiotics are useless at treating the common cold.
6. Interferon treatment prevents about 70 percent of common colds.

Lower Respiratory Infections

A. The following statements are about bacterial infections. Complete each statement with the correct term or terms.
1. Pneumoncoccal pneumonia is caused by the bacterium _____ _____.

2. The swelling of the capsule in a pneumonia-causing bacterium is called a _____ reaction.

3. In the respiratory system, aspirating is the insertion of a needle into the _____ or _____.

4. A consolidated region of the lung is filled with _____.

5. By pneumococcal _____, pneumococci multiply in the blood.

6. _____ is the name for the vaccine used for immunization against pneumoncoccal pneumonia.

7. _____ _____ is the causative agent of mycoplasmal pneumonia.

8. Mycoplasmal pneumonia is usually treated with the antibiotic _____.

9. Chlamydiae are obligate _____ _____.

10. *C. psittaci* commonly affects all kinds of _____.

11. Q fever is caused by the bacterium _____ _____.

12. Q fever is usually treated with the antibiotic _____.

13. *L. pneumophila* is a _____-shaped bacterium.

14. *L. pneumophilia* is _____ to penicillin.

15. Another name for pertussis is _____.

16. *B. pertussis* produces a protein called filamentous _____, which attaches to susceptible cells.

17. _____ is the leading killer among infectious diseases.

18. *Mycobacterium tuberculosis* is detected by the _____ stain.

19. The doubling time for *M. tuberculosis* is about _____ hours.

20. Calcified tubercles caused by *M. tuberculosis* are called _____ complexes.

21. Tuberculosis _____ infects the central nervous system.

22. BCG vaccination interferes with tuberculin _____ _____.

23. During the 1980s the rate of tuberculosis cases has been _____.

24. *M. bovis* infects _____.

25. *M. avium* causes a form of the disease _____.

B. The following statements are about viral infections. Label each statement as true or false. If false, correct it.

1. Infection by the influenza virus stimulates the production of antibodies that prevent reinfection.

2. Influenza viruses are identified by their HA and NA antigens.

3. The influenza virus can be transmitted directly from birds to humans.

4. Bronchiolitis is common in infants.

5. Vaccination protects about 20 percent of recipients from influenza.

6. Amantadine is effective as an immunization in preventing influenza A.

7. Croup is most common in adults.

8. Croup epidemics occur during the fall.

9. RSV infection occurs during the late winter and early spring.

10. The virus for the hantavirus pulmonary syndrome occurs mainly in rats.

C. The following statements are about fungal infections. Label each statement as true or false. If false, correct it.

1. Fungal infections commonly infect the lower respiratory tract.

2. Histoplasmosis occurs mainly in the United States.

3. About 50 percent of the people infected with the histoplasmosis fungus become ill.

4. Most people infected with coccidiodomycosis suffer only a mild respiratory illness.

5. Blastomycosis symptoms can resemble pulmonary tuberculosis.

6. *Pneumocystis carinii* is a protozoan.

Multiple Choice: Review

1. Select the incorrect association.
   A. larynx/voicebox
   B. pleura/lines bronchi
   C. tonsil/lymphoid tissue
   D. trachea/windpipe

2. By bronchitis, the bronchi
   A. shrink and produce a thin mucus.
   B. shrink and produce a thick mucus.
   C. swell and produce a thin mucus.
   D. swell and produce a thick mucus.

3. The lettered groups of *Streptococcus pyrogenes* refer to the type of
   A. Gram-staining reactions of the bacteria.
   B. location of infection in the human body.
   C. polysaccharide antigen present on cell surfaces.
   D. tendency to produce hemolysis in infected cells.

4. Rheumatic fever causes the most life-threatening damage to the
   A. heart.
   B. kidneys.
   C. joints.
   D. skeletal muscles.

5. The diphtheria toxin is a typical
   A. A toxin.
   B. AB toxin.
   C. ABC toxin.
   D. X toxin.

6. The common cold is caused by a

    A. bacterium.
    B. fungus.
    C. protozoan.
    D. virus.

7. The coronavirus is a

    A. large-sized DNA virus.
    B. large-sized RNA virus.
    C. medium-sized DNA virus.
    D. medium-sized RNA virus.

8. About _____ percent of healthy adults harbor *S. pneumoniae* in their throats.

    A. five
    B. ten
    C. twenty
    D. fifty

9. Each is a genus causing pneumonia except

    A. *Haemophilus.*
    B. *Klebsiella.*
    C. *Pseudomonas.*
    D. *Staphylococcus.*

10. Q fever is caused by a

    A. fungus.
    B. protozoan.
    C. rickettsia.
    D. virus.

Answers

Correlation Questions

1. The chemicals and irritants from smoking destroy lines of defense in the upper respiratory tract, namely the mucous membranes and the cilia.

2. The enlargement of the tonsils can block the respiratory tract, inhibiting the free inflow and outflow of air.

3. The eustachian tube is a passageway from the throat to the middle ear. It serves as an avenue to move air from the throat to restore proper air pressure on the tympanum (ear drum). However, it is also an avenue for the migration of microbes from the throat to the middle ear, some which can cause infection.

4. The throat is the pharynx. Treatment should center on the larynx (voice box) of the respiratory tract.

Clinical Science

1. etiology   2. Auscultation   3. anatomical   4. throat   5. persistent

The Respiratory System

A. 1. larynx   2. nasopharynx   3. trachea   4. alveoli   5. bronchioles and alveoli

B. 1. It is warm, moist, and nutrient-rich.
   2. It is nasal inflammation.
   3. It is the infection of the middle ear.
   4. It is rapid breathing.
   5. They become filled with fluid.

Upper Respiratory Infections

A. 1. True
   2. True
   3. False; This antibiotic is almost useless for this treatment.
   4. False; Beta-hemolytic strains lyse red blood cells.
   5. True
   6. False; Most people infected with this recover in a few days.
   7. False; Type A is the most virulent toxin.
   8. True
   9. False; It is an inflammation of the kidney.
   10. False; It is Gram-positive.
   11. True
   12. False; It will only stop further production and transmission of the disease.
   13. True
   14. True

B. 1. True
   2. False; They are single-stranded RNA viruses.
   3. True
   4. False; They are protected for 18 months.
   5. True

6. False; This treatment prevents about 40 percent of common colds.

Lower Respiratory Infections

A. 1. *Streptococcus pneumoniae*
   2. quelling
   3. lungs or trachea
   4. fluid
   5. sepsis
   6. Pneumovax
   7. *Mycoplasma pneumoniae*
   8. tetracycline
   9. intracellular parasites
   10. birds
   11. *Coxiella burnetii*
   12. tetracycline
   13. rod
   14. resistant
   15. whooping cough
   16. hemagglutinin
   17. tuberculosis
   18. acid-fast
   19. twenty
   20. Ghon
   21. meningitis
   22. skin testing
   23. increasing
   24. cows
   25. tuberculosis

B. 1. True
   2. True
   3. False; They can exchange it with pigs that infect humans.
   4. True
   5. False; It protects about 70 percent of the recipients.
   6. True
   7. False; It is most common in toddlers.
   8. True

9. True

10. False; It occurs mainly in long-tailed deer mice.

C. 1. True

2. False; It occurs worldwide.

3. False; Only about one percent of the people infected become ill.

4. True

5. True

6. False; It is a fungus.

Multiple Choice: Review

1. B   2. D   3. C   4. A   5. B   6. D   7. D   8. B   9. C   10. C

Chapter 23
Infections of the Digestive System

The Digestive System
    Structure and Function
    Clinical Syndromes
Infections of the Oral Cavity and Salivary Glands
    Bacterial Infections
    Viral Infections
Infections of the Intestinal Tract
    Bacterial Infections
    Viral Infections
    Protozoal Infections
    Helminth Infections
Infections of the Liver
    Viral Infections
    Helminthic Infections
Summary

Key Terms

alimentary canal	peristalsis
esophagus	chyme
duodenum	jejunum
ileum	large intestine
gallbladder	gastritis
gastroenteritis	hepatitis
cariogenic	periodontium
gingivitis	exanthems
orchitis	Shiga toxin
botulism	infectious hepatitis
hepatocytes	

Study Tips

1. Most of the information in this chapter relates digestive diseases to causative microorganisms. Organize this information by composing the following table.

            Disease   Microorganism

2. Study the anatomy of the digestive system in more depth by consulting texts on human anatomy and physiology. Learning more about these topics will help you to understand the microorganisms and diseases discussed throughout the chapter.

3. Does your biology lab have a human torso model? Study it as another aid to enhance your understanding of the human anatomy relevant to this chapter and other chapters.

4. Study prepared slides of some of the causative microorganisms described in this chapter. Pay particular attention to the cell morphology and Gram-staining reactions described.

5. Conduct an Internet search on the topics of this chapter. To understand some of the concepts, images with motion can be helpful. For the digestive system, start with peristalsis - animation.

Correlation Questions

1. The digestive system has numerous adaptations to facilitate the breakdown and use of nutrient molecules. How is each of the following adaptive: molars, peristalsis, and villi lining the inside surface of the small intestine?

2. Stomach ulcers are caused by a bacterial infection. How can stress contribute to the development of ulcers?

3. How does a change in pH the oral cavity contribute to the development of dental caries?

4. Select a region of the digestive tract such as the large intestine. What characteristics of this region make it a welcome environment for bacterial infection?

The Digestive System

A. Complete each of the following statements with the correct term or terms.
1. _____ is the intestinal movement that propels food through the digestive tract.
2. The secretion of bile is specific for the breakdown of lumps of _____ in the diet.
3. The _____, _____, and _____ are regions of the small intestine.
4. The pancreas does not have a normal biota due to _____ and _____ immune defenses.
5. _____ is the inflammation of the stomach.
6. Colitis mainly affects the _____ of the digestive tract.
7. The term "periodontal" means _____.
8. Hepatitis involves damage to the _____.

Infections of the Oral Cavity and Salivary Glands

A. Complete each of the following statements with the correct term or terms.
1. *S. mutans* attaches to the _____ of the oral cavity.
2. *S. mutans* is _____, meaning that it is caries-producing.
3. The fermentation of sucrose in the oral cavity produces _____ acid.
4. Incorporation of the mineral _____ into tooth enamel makes the teeth more resistant to caries.
5. The root of a tooth is normally covered with _____.

6. _____ is the inflammation of the gums.

7. The primary causative bacterium of gingivitis is _____.

8. _____ are skin rashes.

9. The microorganism that causes mumps is a _____.

10. Since the MMR vaccine the incidence of mumps in the U.S. have decreased _____ percent.

Infections of the Intestinal Tract

A. Label each of the following statements as true or false. If false, correct it.

1. *Shigella* species are Gram-negative.

2. *Shigella* species cause microabscesses in the colon.

3. Shigellosis has been nearly eradicated in the United States.

4. More than a 15 percent loss of human body weight by dehydration is usually fatal.

5. Typhoid fever is caused by a species of *Streptococcus*.

6. Typhoid fever has become a rare disease in developed countries.

7. Salmonellosis is a true infection caused by bacteria multiplying in the intestine.

8. Salmonellosis is usually treated with antibiotics.

9. *E. coli* is an invasive pathogen to the human intestine.

10. The heat labile toxin causing *E. coli* diarrhea is similar to the toxin that causes cholera.

11. Enteroinvasive strains of *E. coli* cause a dysentery almost identical to shigellosis.

12. Cholera is caused by a species of *Vibrio*.

13. Chlorea is caused by the retention of chloride ions in the intestinal tract.

14. The fluids of ORT contain glucose and sodium.

15. *V. parahaemolyticus* is a Gram-positive rod.

16. A species of *Yersinia* causes enterocolitis.

17. *Campylobacter* species are slightly curved, Gram-negative rods.

18. Stomach ulcers are caused by a virus.

19. Iatrogenic diarrhea is caused by a species of *Staphylococcus*.

20. Strains of *S. aureus* that cause food poisoning produce a heat-stable protein enterotoxin.

21. *B. cereus* lives in soil and water.

22. *C. perfringens* causes botulism.

23. *C. botulinum* causes gas gangrene.

24. A rotavirus can cause gastroenteritis.

25. About one-third of all gastroenteritis outbreaks are caused by Norwalk agents.

B. Match each protozoan disease to its correct description.

1. amebic dysentery   A. It causes sever problems for AIDS patients.
2. giardiasis   B. It is caused by a ciliated protozoan.
3. balantidiasis   C. It occurs as a trophozoite.
4. cryptosporidiosis   D. It is caused by a protozoan with the species
                                                    name *histolyticia*.

C. Helminth Infections/Short Answer

1. Which helminth causes the most common helminthic infection?
2. What is the name of the microscopic sites in the lungs where *Ascaris lumbricoides* matures?
3. How do hookworms penetrate the human body?
4. Where do the eggs of *Strongyloides stercoralis* hatch in the human body?
5. What is the genus name for the whipworm?
6. How are humans infected in cysticercosis?

Infections of the Liver

A. Complete each of the following statements with the correct term.

1. The hepatitis A virus contains single-stranded _____.
2. _____ are liver cells.
3. Hepatitis A is distinguished by the presence of _____ antibodies.
4. The _____ virus is the best-known virus to cause human cancer.
5. Worldwide there are more than _____ million carriers of HBV.
6. A new _____ could eradicate hepatitis B from the human population.
7. HCV is transmitted by sexual contact or _____.
8. Chronic hepatitis delta can progress to the disease _____ of the liver.
9. The nucleic acid in HEV is _____.
10. *Fasciola hepatica* is the _____ liver fluke.

Multiple Choice: Review

1. Bile is made by the __1__ and stored mainly in the __2__.
   A. 1 - gallbladder, 2 - liver
   B. 1 - liver, 2 - gallbladder

2. Dental caries is caused by a species of the genus

    A. *Bacillus.*
    B. *Escherichia.*
    C. *Staphylococcus.*
    D. *Streptococcus.*

3. Gingivitis is caused by a species of the genus

    A. *Bacillus.*
    B. *Porphyromonas.*
    C. *Staphylococcus.*
    D. *Streptococcus.*

4. Shigellosis is caused by a

    A. Gram-negative coccus.
    B. Gram-negative rod.
    C. Gram-positive coccus.
    D. Gram-positive rod.

5. During dehydration, in the intestinal tract

    A. chloride ions follow the loss of water.
    B. water follows the loss of chloride ions.

6. Select the correct statement about salmonellosis.

    A. It is a true infection.
    B. It is no longer a major health concern.
    C. Most healthy adults with it recover in a few days.
    D. It is usually not self-limiting.

7. Diarrhea caused by *E. coli* is treated best with the antibiotic

    A. cephalosporin.
    B. ciprofloxacin.
    C. erythromycin.
    D. tetracycline.

8. Cholera is caused by a species of the genus

    A. *Clostridium.*
    B. *Pseudomonas.*
    C. *Staphylococcus.*
    D. *Vibrio.*

9. Giardiasis is caused by a

    A. bacterium.
    B. fungus.
    C. protozoan.
    D. virus.

10. Select the cancer-causing virus.
    A. HAV
    B. HBV
    C. HCV
    D. HEV

Answers

Correlation Questions

1. Molars are teeth providing broad, flat surfaces for crushing and grinding food. Peristalsis provides the forces to break down and mix food physically as this action moves food through the stomach, small intestine, and large intestine. The villi increase the surface area in the small intestine as it contacts food molecules, facilitating their absorption from the digestive tract into the blood.

2. Stress can weaken the immune system, making the human body more vulnerable to the attack of any invasive microbe.

3. The metabolism of some bacteria produces acids that can erode away the enamel of the teeth.

4. Favorable factors include a warm, moist, and dark environment.

The Digestive System

A. 1. peristalsis
   2. fat
   3. duodenum, jejunum, ileum
   4. innate, adaptive
   5. gastritis
   6. colon
   7. around the teeth
   8. liver

Infections of the Oral Cavity and Salivary Glands

A. 1. teeth
   2. cariogenic
   3. lactic
   4. fluoride
   5. cementum

6. gingivitis
7. *Porphyromonas gingivalis*
8. enanthems
9. virus
10. 97

Infections of the Intestinal Tract

A. 1. False; They are Gram-positive.
   2. True
   3. False; Thousands of cases are reported annually.
   4. True
   5. False; It is caused by *Salmonella typhi.*
   6. True
   7. True
   8. False; This is contraindicated.
   9. False; It is a normal biota of the human intestine.
   10. True
   11. True
   12. True
   13. False; It is stimulated by the loss of chloride ions from the intestinal tract.
   14. True
   15. False; It is Gram-negative.
   16. True
   17. True
   18. False; They are caused by a bacterium.
   19. False; It is medically induced diarrhea cause by *C. difficile.*
   20. True
   21. True
   22. False; *C. botulinum* causes botulism.
   23. False; It causes botulism.
   24. True
   25. True

B. 1. D 2. C 3. B 4. A

C. 1. pinworm
   2. alveoli

3. They penetrate the human skin.
4. human intestine
5. *Trichuris*
6. Larvae encyst in human tissues.

Infections of the Liver

A. Complete each of the following statements with the correct term.
1. RNA
2. hepatocytes
3. antiviral
4. HBV
5. 300
6. vaccine
7. contaminated blood
8. cirrhosis
9. RNA
10. sheep

Multiple Choice: Review

1. B   2. D   3. B   4. B   5. B   6. A   7. B   8. D   9. C   10. B

Chapter 24
Infections of the Genitourinary System

Key Terms

glomerulus

urethra

gamete

prostate gland

uterus

vagina

sexually transmissible disease

urethritis

urinary tract infection

salpingitis

pelvic inflammatory disease

leukocyte esterase

gonococcus

purulent

snuffles

granuloma inguinale

puerperal sepsis

bladder

semen

testes

ovary

cervix

embryo

cystitis

pyelonephritis

vaginitis

oophoritis

perinatal

kidney stone

gonorrheal endotoxin

syphilis

lymphogranuloma venereum

primary infection

Study Tips

1. Much of the information in this chapter relates genitourinary diseases to causative microorganisms. Organize this information by composing the following table:

              Disease   Microorganism

2. Consult additional sources to study the anatomy and physiology of the urinary and reproductive systems in more depth. This will promote your understanding of the infections of the urinary and reproductive tracts.

3. Your lab experiences can also facilitate understanding of the concepts in this chapter. Are there models available in your lab showing the urinary and reproductive systems? Have you studied some of the microorganisms described in this chapter? Start by reviewing their cell morphology and Gram-stain reactions.

4. Continue to answer the questions at the end of the chapter. After answering these questions independently, ask your instructor to post an answer key for these questions.

Correlation Questions

1. How can an HSV infection lead to the loss of motor control in the human body?

2. What conditions in the female reproductive tract are favorable for the growth of microbes?

3. How is the development of dysuria related to the infection of the urinary tract?

4. How is a knowledge of the blood important to understand the diseases of the urinary and reproductive systems?

The Genitourinary System/Structure and Function of the Urinary System

A. Complete each of the following statements with the correct term or terms.
   1. The _____ and _____ tracts merge in the male.
   2. The glomerulus is a cluster of _____ _____ in the kidney.
   3. Urine leaves the kidney through the _____ of the urinary tract.
   4. Urine is stored in the _____.
   5. In the male, the _____ of the urinary tract is a longer structure than in the female.

Structure and Function of the Reproductive System

A. Complete each of the following statements with the correct term or terms.
   1. A _____ is an ovum or a sperm.
   2. In the male the _____ is carried in the scrotum.

3. _____ is a mixture of sperm cells and seminal fluid.
4. The _____ is a neck-like extension of the uterus.
5. The internal lining of the urinary tract consists of tightly-joined _____ cells.
6. _____ is the infection of the bladder.
7. _____ is an infection of the lining of uterus.
8. _____ is an infection of the fallopian tubes.

Urinary Tract Infections

A. Label each of the following statements as true or false. If false, correct it.
   1. All major pathogens of the urinary tract are bacteria.
   2. Urinary tract infections are extremely common.
   3. Males are more vulnerable than females to UTIs.
   4. A normal urine sample contains very few bacteria.
   5. The presence of leukocyte esterase in the urine indicates the breakdown of white blood cells in the urine.
   6. *Staphylococcus aureus* is the most common pathogen of the urinary tract.
   7. *Leptospira interrogans* usually enters the human body by a break in the serous membranes of the skin.
   8. *L. interrogans* is a large bacterium and easy to diagnose.

Sexually Transmissible Diseases (STDs)

A. Label each of the following statements as true or false. If false, correct it.
   1. Gonorrhea is caused by a species of *Neisseria*.
   2. The rate of outbreaks of gonorrhea has continued to increase over the last 20 years.
   3. The presence of Gram-positive diplococci in an urethral discharge indicates the presence of *N. gonorrhoeae* in the urinary tract.
   4. The presence of capsules on gonococcal cells promotes adherence of these pathogens to the epithelial of the urinary tract.
   5. *N. gonorrhoeae* infects only humans.
   6. Gonococcal infections can lead to the development of arthritis.
   7. Gonococci can become penicillin-resistant through chromosomal mutations or by acquiring a plasmid.
   8. Recently, in the United States the rate of syphilis outbreaks has increased markedly.
   9. *T. pallidum* infects only human beings.
   10. Secondary syphilis begins 12 to 14 weeks after the chancre appears.
   11. Latent syphilis develops after late syphilis.

12. Gummas develop in the cardiovascular system.
13. Chlamydia is the most prevalent STD worldwide.
14. Chlamydia can be cured with erythromycin.
15. Lymphogranuloma venereum develops in the cervical lymph nodes.
16. *C. granulomatis* is a Gram-negative rod.
17. Herpes simplex is the most commonly sexually transmitted pathogen in industrialized countries.
18. During its active stage, HSV multiplies rapidly.
19. The initial infection by HSV is the primary infection.
20. The administration of acyclovir can cure genital herpes.
21. Genital warts is commonly reported.
22. HPV can lead to cancer.

Infections of the Female Reproductive Tract

A. Complete each of the following with the correct term.
   1. *G. vaginalis* is a Gram _____ bacterium.
   2. *G. vaginitis* can be diagnosed if _____ cells are sloughed off from the vagina.
   3. Toxic shock syndrome is caused by strains of _____ _____.
   4. PID increases the risk for a pregnancy that is _____.
   5. Endometritis is particularly likely to occur after _____ or _____.
   6. _____ streptococci are one causative group for endometritis.
   7. *Candida albicans* is a yeast-like _____.
   8. *T. vaginalis* is a _____ protozoan.

Infections of the Male Reproductive Tract

A. Complete each of the following with the correct term.
   1. _____ is the inflammation of the male reproductive tract.
   2. _____ _____ is the most common causative agent of nongonococcal urethritis in males.

Infections Transmitted from Mother to Infant

A. Label whether each of the following is caused by a bacterium or a virus.
   1. cytomegalic inclusion disease
   2. listeriosis

B. Complete each of the following with the correct term or terms.
   1. *L. monocytogenes* is an opportunistic _____.

2. Group B streptococci is found in the _____, _____, and vagina.

3. CMV is transmitted by the exchange of _____ and _____.

4. _____ is an effective drug for treating CMV.

Multiple Choice: Review

1. Infection of the fallopian tubes is

   A. endometritis.
   B. oophoritis.
   C. salpingitis.
   D. vaginitis.

2. The pathogen most commonly infecting the female reproductive tract is

   A. *B. subtilis.*
   B. *E. coli.*
   C. *P. aeruginosa.*
   D. *S. aureus.*

3. 90 percent of *E. coli* strains are sensitive to

   A. cefalexin.
   B. erythromycin.
   C. streptomycin.
   D. tetracyline.

4. Select the incorrect statement about *L. interrogans.*

   A. It is a member of the spirochete family.
   B. It is a small bacterium.
   C. It is difficult to diagnose.
   D. It is reported frequently in the U.S.

5. Select the incorrect statement about *N. gonorrhoeae.*

   A. It is a fragile pathogen.
   B. It is a highly adapted pathogen.
   C. It is easily killed when exposed to sunlight.
   D. It is Gram-positive.

6. Skin lesions appear during _____ syphilis.

   A. primary
   B. secondary
   C. tertiary
   D. quaternary

7. What is the most prevalent STD worldwide?

   A. chlamydia
   B. gonorrhea
   C. herpes
   D. syphilis

8. HSV is a(n)

   A. DNA virus that affects the heart.
   B. DNA virus that affects neurons.
   C. RNA virus that affects the heart.
   D. RNA virus that affects neurons.

9. Genital warts is caused by a

   A. bacterium.
   B. fungus.
   C. protozoan.
   D. virus.

10. TSST is caused by a species of

   A. *Escherichia.*
   B. *Pseudomonas.*
   C. *Staphylococcus.*
   D. *Streptococcus.*

Answers

Correlation Questions

1. The HSV infects motor neurons that signal the skeletal muscles of the body. Impairment of this signaling system can hinder body movements through the loss of skeletal muscle contraction.

2. Favorable conditions include a warm and moist environment with the absence of light.

3. The infection from a microbe can cause inflammation of the mucous membranes of the urinary tract. One of the symptoms of inflammation is pain.

4. The immune responses of the body are reflected in the changing composition of the blood. Examples include the production of an enzyme such as leukocyte esterase or the changing percentages of white blood cells.

The Genotiurinary System/Structure and Function of the Reproductive System

A. 1. urinary, reproductive
   2. blood vessels

3. ureter
4. bladder
5. urethra

Structure and Function of the Reproductive System

A. 1. gamete
2. testis
3. semen
4. cervix
5. epithelial
6. cystitis
7. endometritis
8. salpingitis

Urinary Tract Infections

A. 1. True
2. True
3. False; Females are more vulnerable.
4. True
5. False; It indicates that white blood cells are present in the urine.
6. False; *E. coli* is the most common pathogen of the urinary tract.
7. False; It enters by a break in the mucous membranes.
8. False; It is small and difficult to detect.

Sexually Transmissible Diseases (STD's)

A. 1. True
2. False; It had declined steadily but had an abrupt increase in 1998.
3. False; Finding Gram-negative diplococci indicates its presence.
4. False; Pili promote the adherence of this microbe.
5. True
6. True
7. True
8. False; It has declined steadily over the last ten years.
9. True
10. False; It occurs 6 to 8 weeks after a chancre appears.

11. False; It develops before late syphilis.

12. False; They develop in bones or skin.

13. True

14. True

15. False; It develops in the lymph nodes of the groin.

16. False; It is a Gram-positive rod.

17. True

18. True

19. True

20. False; It cannot be cured.

21. False; It is not a reportable disease.

22. True

Infections of the Female Reproductive Tract

A. 1. positive
    2. clue
    3. *Staphylococcus aureus*
    4. ectopic
    5. childbirth, abortion
    6. anaerobic
    7. fungus
    8. flagellated

Infections of the Male Reproductive Tract

A. 1. urethritis
    2. *Chlamydia trachomatis*

Infections Transmitted from Mother to Infant

A. 1. virus
    2. bacterium

B. 1. pathogen
    2. throat, gastrointestinal tract
    3. saliva and infected blood
    4. gangiclovir

Multiple Choice: Review

1. C   2. B   3. A   4. D   5. D   6. B   7. A   8. B   9 D   10. C

Chapter 25
Infections of the Nervous System

The Nervous System
    Structure and Function
    Clinical Syndromes
Infections of the Meninges
    Bacterial Causes
    Viral Causes: Aseptic Meningitis
    Fungal Causes
Diseases of Neural Tissue
    Bacterial Causes
    Viral Causes
    Prion Causes
    Protozoal Causes
Summary

Key Terms

central nervous system

nerve

dura mater

pia mater

lumbar puncture

meningitis

myelitis

intoxication

invasive

aseptic meningitis

tetanus

trismus

botulism

rabies

excitation phase

paralytic phase

subclinical

peripheral nervous system

meninges

arachnoid

cerebrospinal fluid

blood-brain barrier

encephalitis

neurotoxin

protease

meningococcal prophylaxis

chronic meningitis

tetanospasmin

tetanus antitoxin

botulinum toxin

prodromal phase

hydrophobia

arbovirus

African sleeping sickness

Study Tips

1. Answer the several kinds of questions at the end of the chapter. After answering them, discuss your answers with your instructor and classmates.

2. New vocabulary is a major part of this chapter. Organize the key terms as follows:

   Term   Definition

   After studying the text, define each term in your own words.

3. Can you tie in information from other courses to the content of this chapter? Start by outlining the human central and peripheral nervous system. Describe the functions of the neuron and neuroglial cells. What is the composition of the blood? What is the structure and function of the respiratory system?

4. These concluding chapters in the text draw on basic information that you have already studied. Can you find this information in the text? Begin with basic chemistry (Chapter 2), prokaryotic cells (Chapter 4), and viruses (Chapter 13). A review of these chapters will help you to understand the content of Chapter 25.

5. Consult the Internet to expand your knowledge on the topics in this chapter. Use the keys terms from the chapter as search terms.

Correlation Questions

1. The nervous system consists of two branches: the central nervous system and the peripheral nervous system. Explain how these two branches work together.

2. How does the blood brain barrier make many infections of the brain difficult to treat?

3. What do you think is the difference among the following: sensory nerve, motor nerve, and mixed nerve?

4. Can meningitis and encephalitis be distinguished on the basis of brain function?

The Nervous System

A. Label each of the following statements as true or false. If false, correct it.
   1. Neuroglia cells send impulses.
   2. The spinal cord is part of the CNS.
   3. The dura mater is the innermost meningeal layer.
   4. The CSF circulates between the dura mater and arachnoid mater.
   5. Glucose can pass through the blood brain barrier.
   6. Some antibiotics cannot pass through the blood brain barrier.
   7. Myelitis is the inflammation of the meninges.
   8. Encephalitis is inflammation of brain tissue.

Infections of the Meninges: Bacterial Causes

A. Select the correct statements about *Neisseria meningitidis*.
1. It is a Gram-positive pathogen.
2. Its only natural reservoir is the human body.
3. It is usually grown on a very rich nutrient medium.
4. It is transmitted from person to person by respiratory droplets.
5. It adds iron to transferrin in the human body.
6. One of the symptoms of its disease is producing petechiae on the body.
7. It cannot be treated with antibiotics.
8. A vaccine is now available to treat the meningitis it causes.

B. Select the correct statements about *Haemophilus influenzae*.
1. It has a polysaccharide capsule.
2. It infects the urinary tract.
3. Tetracycline is commonly prescribed to treat carriers of this microbe.
4. Until 10 years ago it caused 75 percent of all meningitis in infants and young children.

C. Select the correct statements about *Streptococcus pneumoniae*.
1. It is an encapsulated pathogen.
2. It causes 90 percent of the cases of meningitis in adults over age 40.
3. Alcoholics are very vulnerable to its infection.
4. It enters the CNS from the digestive tract.

D. Select the correct statement about *E. coli*.
1. It causes less than 5 percent of the bacterial meningitis cases.
2. 90 percent of patients with meningitis, and infected with it, are cured with the antibiotics ampicillin or gentamicin.

Infections of the Meninges: Viral and Fungal Causes

A. Select the correct statements about aseptic meningitis.
1. It is fairly common.
2. It is caused by one specific virus.
3. Antimicrobial treatment is effective against it.
4. It is a deadly disease.
5. It should be treated with antibiotics until bacterial meningitis is ruled out.

B. Select the correct statements about *Cryptococcus neoformans*.
1. It reduces the leukocyte count per cubic mm.
2. It is a bacterium.
3. It is dimorphic.
4. It is inhaled into the lungs.

Diseases of Neural Tissues: Bacterial Causes

A. Complete each of the following statements about *Clostridium tetani* with the correct term.
1. It is a Gram-_____ rod.
2. It moves by _____ flagella.
3. This bacterium produces _____ under harsh conditions.
4. It grows best at a temperature of _____ degrees Celsius.
5. It grows best at a pH of _____.
6. The wound it infects becomes _____ when the blood supply to it is cut off.
7. It produces a(n) _____, which has a deadly effect on the nervous system.
8. Tetanospasmin affects neurons that stimulate _____.
9. _____ is another name for lockjaw.
10. The treatment for lockjaw is the injection of _____.

B. Complete each of the following statements about *Clostridium botulinum* with the correct term.
1. Its cells are _____-shaped.
2. It produces heat-resistant _____.
3. The botulinum toxin is designated A through _____.
4. Type _____ toxin is the most potent toxin.
5. Its toxin prevents the release of _____ from neurons.
6. It produces _____ paralysis in muscles.
7. Most adults acquire botulism from _____.
8. The process of _____ can destroy its toxin.

Diseases of Neural Tissue: Viral Causes/Prion Causes

A. Select the correct statements about the rabies virus.
1. It contains a single minus strand of RNA.
2. It is usually transmitted by an animal bite.
3. Its earliest phase of rabies is the prodromal phase.
4. The excitation phase is the last phase of rabies.

5. Pasteur's treatment is still used commonly to treat this disease.

B. Select the correct statements about polio and the polio virus.
 1. The virus is a DNA virus.
 2. The virus has seven different serotypes.
 3. Cases of polio were reported in the Western Hemisphere in 1996.
 4. OPV is the Sabin vaccine.
 5. IPV is the Salk vaccine.
 6. Polio has been eradicated in the United States.

C. Select the correct statements about the arboviruses.
 1. They are a diverse group of RNA viruses.
 2. The different types of encephalitis are caused by one arbovirus.
 3. All of the viruses are transmitted by mosquitoes.
 4. Diagnosis of the encephalitis caused by an arbovirus is difficult.
 5. The West Nile virus is an arbovirus.

D. Select the correct statements about the prions and the diseases they cause.
 1. The various prion-caused diseases have the same mode of infection.
 2. CJD is not a hereditary disease.
 3. At least six prion diseases afflict animals.
 4. One prion disease is BSE.

Diseases of Neural Tissue: Protozoal Causes

A. Select the correct statements about trypanosome diseases.
 1. Two subspecies cause African sleeping sickness.
 2. The trypanosomes are transmitted to humans by the bite of the tsetse fly.
 3. The trypanosomes damage the brain tissue.
 4. Sleeping sickness is a major cause of death in Africa.

Multiple Choice: Review

1. The innermost layer of the meninges is the
 A. arachnoid mater.
 B. cerebral cortex.
 C. dura mater.
 D. pia mater.

2. Encephalitis is an inflammation of the

   A. brain.
   B. heart.
   C. kidney.
   D. skeletal muscles.

3. Select the incorrect statement about *N. meningitidis*.

   A. It avoids the defenses of the human body.
   B. It forms endospores.
   C. It grows well in chocolate agar.
   D. It is Gram-negative.

4. *C. neoformans* enters the body by

   A. breakage in the skin.
   B. contaminated food.
   C. inhalation.
   D. urinary tract infections.

5. Tetanospasmin is produced by a species of

   A. *Bacillus.*
   B. *Clostridium.*
   C. *Escherichia.*
   D. *Streptococcus.*

6. Which strains of *C. botulinum* cause human disease?

   A. A, B, C
   B. A, B, E
   C. C, D, E
   D. D, E, F

7. Select the incorrect statement about the rabies virus.

   A. It belongs to the rhabdovirus family.
   B. It infects many different animals.
   C. It is a DNA virus.
   D. It is bullet-shaped.

8. The Pasteur treatment involves injections into the _____ wall.

   A. abdominal
   B. pericardial
   C. periosteal
   D. pleural

9. The Sabin vaccine is the
   A. IPV.
   B. OPV.

10. African sleeping sickness is caused by the infection of a
    A. bacterium.
    B. fungus.
    C. protozoan.
    D. virus.

Answers

Correlation Questions

1. The central nervous system is the central control center. The brain, for example, can receive and sends messages and store information. The spinal cord can coordinate spinal reflexes. Peripherally, the PNS sends signals into central control (CNS) and sends signals away from central control to organs and skeletal muscles.

2. Many antibiotics cannot pass through the blood brain barrier. Due to this impermeability they cannot pass into brain cells where infections are developing.

3. Sensory nerves, with their sensory neurons, send signals in peripheral nerves toward the CNS. Motor nerves, with their motor neurons, send signals in the opposite direction. Mixed nerves contain sensory and motor neurons.

4. Encephalitis affects brain function. Meningitis affects the coverings around the brain. Symptoms are different for the diseases affecting each specific anatomical region.

The Nervous System

A. 1. False; Neuroglial cells support neurons and bind them together.
   2. True
   3. False; The dura mater is the outermost meningeal layer.
   4. False; The CSF circulates between the arachnoid and pia mater.
   5. True
   6. True
   7. False; Myelitis is an inflammation of the spinal cord.
   8. True

Infections of the Meninges: Bacterial Causes

    A. 2, 3, 4, 6, 8
    B. 1, 4
    C. 1, 3
    D. 1

Infections of the Meninges: Viral and Fungal Causes

    A. 1, 5
    B. 1, 4

Diseases of Neural Tissues: Bacterial Causes

A. 1. positive 2. peritrichous 3. endospores 4. thirty-seven 5. seven point four 6. anaerobic 7. endotoxin 8. muscle contraction 9. trismus 10. tetanus immune globulin

B. 1. rod 2. endospores 3. G 4. A 5. acetylcholine 6. flaccid 7. food poisoning 8. heat

Diseases of Neural Tissues: Viral Causes/Prion Causes

    A. 1, 2, 3
    B. 4, 5, 6
    C. 1, 3, 4
    D. 3, 4

Diseases of Neural Tissues: Protozoal Causes

A. 1, 2, 3, 4

Multiple Choice: Review

1. D 2. A 3. B 4. C 5. B 6. B 7. C 8. A 9. B 10. C

Chapter 26
Infections of the Body's Surfaces

The Body's Surfaces
    Structure and Function of the Skin
    Structure and Function of the Eye's Surface
    Defenses and Normal Microbiota: A Brief Review
    Clinical Syndromes
Skin Infections
    Bacterial Causes
    Viral Causes
    Fungal Causes
    Arthropod Causes
Eye Infections
    Bacterial Causes
    Helminthic Causes
Summary

Key Terms

epidermis
erysipelas
leukocidin
scalded skin syndrome
carbuncle
cellulites
gas gangrene
isotretinoin
leprosy
spinal ganglia
dermatome
herpes keratitis
vaccinia
candidiasis
trachoma

impetigo
streptococcal gangrene
exfoliative toxin
folliculitis
abscess
cystic fibrosis
crepitance
teratogen
Hansen's disease
Reye's syndrome
herpetic whitlow
Koplik spots
tinea
scabies
loaiasis

Study Tips

1. Information is always more meaningful if you can organize it as you study it. For most of the information in this chapter, try organizing it by this table:

           Diseases   Microorganisms

2. Add to your knowledge of the skin by studying it through additional texts. The free surfaces of the human body are covered by four different kinds of membranes: cutaneous (skin), synovial, serous, and mucous.

3. Are there resources in your biology lab that will add to your study of the concepts of this chapter? Possibilities include studying a model of the skin, a prepared slide of the skin under the microscope, and a model of the eye.

Correlation Questions

1. An invertebrate animal evolves a skin that consists of a single layer of epithelial cells. Do you think this is as adaptive as a skin with a stratified makeup of epithelial cells.

2. The outer layer of the skin epidermis consists of dead, keratinized cells. What conditions promote their death at this location in the body?

3. How do you think the blood flow through the dermis regulates body temperature?

4. An infecting bacterium produces a hyaluronidase in the body. How does this threaten the health of the body?

The Body's Surfaces

A. Complete each of the following statements with the correct term.
1. The _____ is the outermost layer of the skin.
2. _____ is a waterproof protein of the skin.
3. The _____ is the layer of the skin containing blood vessels.
4. _____ is an enzyme from the skin that kills Gram-positive bacteria.
5. The surface of the eye exposed to the environment is covered with a membrane called the _____.
6. Most of the bacteria living on the skin are not pathogens but are _____.
7. _____ is an infection of the cornea.

Skin Infections

A. The following statements are about skin infections caused by bacteria. Label each statement as true or false. If false, correct it.
1. The most common kind of impetigo affects the deeper layers of the dermis.
2. Impetigo is caused by two different viruses.
3. Erysipelas affects only the superficial layers of the skin.
4. Leukocidins destroy red blood cells.
5. Streptolysins destroy white blood cells.
6. *S. pyrogenes* infections can be successfully treated with penicillin.
7. The exfoliate toxin causes the layers of the skin to separate and peel.
8. Cellulitis is a diffuse, extensive infection.

9. Penicillinase-producing strains of *S. aureus* produce penicillin.

10. *P. aeruginosa* is an opportunistic pathogen.

11. *P. aeroginosa* causes cystic fibrosis in children.

12. Otitis media is swimmer's ear.

13. *C. perfringens* is an anaerobic, Gram-positive bacterium.

14. Acne is caused by the suppression of an inflammatory response.

15. The tetracyclines are very effective at treating acne.

16. *M. leprae* grows best in the warmer regions of the body.

17. Indeterminate leprosy usually develops three to five years after infection.

18. The rate of leprosy in the U.S. is on the rise.

B. The following statements are about skin infections caused by viruses. Label each statement as true or false. If false, correct it.

1. VZV infects only humans.

2. Varicella develops along the dermatomes of the body.

3. The Oka strain has been developed to treat smallpox.

4. Acyclovir is used routinely to treat primary genital herpes.

5. The protein projections off the lipid develop of the rubeola virus allow it to agglutinate red blood cells.

6. Measles is caused by a virus.

7. Measles is a highly communicable disease.

8. There is no treatment for SSPE.

9. Rubella is more communicable than measles or chickenpox.

10. The smallpox virus is a small, RNA virus.

C. The following statements are about skin infections caused by fungi. Label each statement as true or false. If false, correct it.

1. Warts is caused by a fungus.

2. The fungi causing ringworm produce an enzyme that destroys keratin.

3. Griseofulvin specifically targets the dermis of the skin.

4. *Candida* species are opportunistic pathogens.

5. *Candida* species thrive in a dry environment.

6. Scabies is confined mainly to the United States.

7. Scabies is a harmless infection.

8. Lice are usually transferred by direct bodily contact or by fomites.

Eye Infections

A. Short Answer

   1. How is the exposed surface of the eye protected?

   2. Which bacterium is the most usual cause of conjunctivitis?

   3. When trachoma develops in the eye, where do the bacteria normally multiply?

   4. What is the treatment for a scarred cornea?

   5. How is blindness from *N. gonorrhoeae* prevented in a newborn?

   6. Which viruses commonly infect the conjunctiva of the eye?

   7. What is herpetic keratitis?

   8. How is river blindness transmitted?

   9. What is suramin?

  10. Where does loaiasis occur throughout the world?

Multiple Choice: Review

1. Select the incorrect association.

   A. conjunctiva/eye
   B. impetigo/digestive tract
   C. keratitis/cornea
   D. stratum corneum/outer layer

2. Select the incorrect statement about *S. pyrogenes*.

   A. It cannot cause food poisoning.
   B. It is Gram-positive.
   C. It is susceptible to small doses of penicillin.
   D. It infects the skin.

3. Burns are most commonly infected by a species of

   A. *Bacillus.*
   B. *Clostridium.*
   C. *Pseudomonas.*
   D. *Streptococcus.*

4. Select the incorrect statement about *Clostridium perfringens*.

   A. It is abundant in the soil.
   B. It is anaerobic.
   C. It is destroyed by necrotic tissue.
   D. It is Gram-positive.

5. Leprosy is caused by a species of
   A. *Escherichia.*
   B. *Mycobacterium.*
   C. *Staphylococcus.*
   D. *Streptococcus.*

6. Chickenpox is caused by a
   A. bacterium.
   B. fungus.
   C. protozoan.
   D. virus.

7. Dermatomes correspond to the distribution of _____ in the body.
   A. blood vessels
   B. brain lobes
   C. fat storage areas
   D. sensory nerves

8. Measles is caused by a
   A. bacterium.
   B. virus.

9. Which virus is the most powerful teratogen?
   A. measles
   B. mumps
   C. rubella
   D. smallpox

10. Ringworm is caused by a
    A. bacterium.
    B. fungus.
    C. helminth.
    D. virus.

Answers

Correlation Questions

1. A single layer of cells is not thick enough to provide the protection that a stratified (many-layered) arrangement of epithelial cells can offer.

2. These surface cells are too far removed from the blood supply in the underlying dermis. Therefore, they do not have access to glucose, other nutrients, and oxygen. Facing the drying action of the air, they are shed from the body surface.

3. Blood vessels that vasodilate in the dermis can bring blood and heat to the surface, allowing it to escape from the body. Vessels that vasoconstrict keep heat in the body, preventing its escape.

4. This breaks the bonding holding cells together, impairing their ability to work together in the tissues of the human body.

The Body's Surfaces

A. 1. epidermis
    2. keratin
    3. dermis
    4. lysozyme
    5. conjunctiva
    6. commensals
    7. keratitis

Skin Infections

A. 1. False; It is a superficial infection.
    2. False; It is caused by *Staphylococcus aureus* and *Staphylococcus pyrogenes*. These bacteria occur in clusters (staphylococcus).
    3. False; It affects the dermis.
    4. False; They destroy white blood cells.
    5. False; They destroy red blood cells.
    6. True
    7. True
    8. True
    9. False; They produce an enzyme against penicillin.
    10. True
    11. False; It infects children who have developed cystic fibrosis.
    12. False; Otitis externa is swimmer's ear.
    13. True
    14. False; It is caused by an inflammatory response.
    15. True
    16. False; It grows in the cooler regions of the body.
    17. True
    18. False; It is rare but is increasing in frequency.

B. 1. True
    2. True

3. False; It is a vaccine to combat varicella zoster.
4. True.
5. True
6. False; It is fought by cell-mediated immunity through T lymphocytes.
7. True
8. True
9. False; It is not as communicable as these diseases.
10. False; It is a large, double-stranded DNA virus.

C. 1. False; Warts is caused by a virus.
2. True
3. False; It concentrates on the stratum corneum.
4. False; They are commensal fungi.
5. False; They thrive in moist environments.
6. False; It is common worldwide.
7. True
8. True

Eye Infections

A. 1. It is protected by the tear fluid.
2. The bacterium is *Pseudomonas aeruginosa*.
3. They multiply on the conjunctiva.
4. The answer is a corneal transplant.
5. It is treated by the administration of erythromycin.
6. The viral groups are the adenoviruses and enteroviruses.
7. It is an ulceration of the cornea.
8. It is transmitted by black flies and buffalo gnats of *Simulium*.
9. It is an anti-helminth drug used to treat river blindness.
10. It occurs only in Africa.

Multiple Choice: Review

1. B 2. A 3. C 4. C 5. B 6. D 7. D 8. B 9. C 10. B

Chapter 27
Systemic Infections

The Cardiovascular and Lymphatic Systems
    Structure and Function
    Defenses and Normal Biota: A Brief Review
    Clinical Syndromes
Infections of the Heart
    Endocarditis
    Myocarditis
    Pericarditis
Systemic Infections
    Bacterial Infections
    Viral Infections
    Protozoal Infections
    Helminthic Infections
Summary

Key Terms

atrium
valve
systemic circulation
artery
vein
myocarditis
septicemia
transovarially
cutaneous anthrax
infectious mononucleosis
viral load
miracidia
elephantiasis

ventricle
pulmonary circulation
aorta
capillary
endocarditis
pericarditis
bubonic plague
ulceroglandular
anthrax toxin
Burkitt's lymphoma
blackwater fever
cercaria

Study Tips

1. Much of the information in this chapter relates infections of the cardiovascular and lymphatic systems to causative microorganisms. Organize this information by composing the following table:

    Disease   Microorganism

2. Consult additional sources to study the anatomy and physiology of the cardiovascular and lymphatic systems in more depth. This will promote your understanding of the infections of these systems.

3. Is a human torso model available in your lab? Is a model of the heart also available? Use these as aids to study the human anatomy described in this chapter.

4. Continue to answer the questions at the end of the chapter in the text. After answering them, ask your instructor to provide a key for the answers.

5. Expand your knowledge of the cardiovascular system using the Internet. For heart action start with the following search terms: cardiac cycle - animation.

Correlation Questions

1. A tendency to produce high amounts of cholesterol is a hereditary condition, caused by a dominant gene. Two parents have the genotypes Hh and Hh. What is the probability that they will pass on this hereditary tendency to an offspring?

2. The thickness of the myocardium in the two ventricles of the heart is not equal. Explain.

3. How does the thin-walled makeup of capillaries facilitate their role in the body?

4. Why are systemic infections difficult to diagnose?

The Cardiovascular and Lymphatic Systems

A. Complete each of the following statements with the correct term or terms.
   1. A heart is dissected in a biology lab. One of the chambers examined has a thick wall. The largest vessel attached to the heart attaches to this chamber. The chamber is the _____ _____.
   2. The action of the valves prevents the _____ of blood through the heart.
   3. Blood returning to the heart from the lungs has a comparatively higher concentration of the gas _____. It has a comparatively lover concentration of the gas _____.
   4. A large blood vessel with a comparatively thin wall is discovered in the dissection of a mammal. This blood vessel is a _____.
   5. The _____ are the microscopic blood vessels of the circulatory system.
   6. In one example of a disease, fluid collects around the heart. This fluid is trapped by the _____ membrane.

Infections of the Heart

A. Complete each of the following statements with the correct terms or terms.
   1. A bacterium adheres to the inside free surface of the heart. It is attaching to the _____ of the heart.
   2. _____ are bulky masses of bacteria and clots.
   3. Among intravenous drug users with acute bacterial endocarditis, the bacterium _____ _____ causes over one-half of the cases of endocarditis.

4. In the U.S. most of the cases of endocarditis are caused by _____ microorganisms.

5. Another name for trypanosomiasis is _____ disease.

6. Cells vulnerable to trypanosomiasis include brain cells, _____ cells, and _____ cells.

7. _____ are the most common cause of viral endocarditis.

8. _____ is the surgical removal of the pericardium.

Systemic Infections

A. The following statements are about bacterial infections. Label each statement as true or false. If false, correct it.

1. Humans are the main reservoir for *Yersinia pestis*.

2. To cause the plague, *Yersinia pestis* enters the human lymphatic system.

3. The bubonic plague is more virulent than the pneumonic plague.

4. Tularemia only occasionally infects humans.

5. Bacteria causing tularemia are killed in the human body by phagocytosis.

6. Species of *Brucella* are Gram-negative coccobacilli.

7. Species of *Brucella* are able to survive as intracellular parasites in phagocytes.

8. Arthritis develops is a person within one month after being bitten by the tick that spreads Lyme disease.

9. Lyme disease can be eradicated from a geographical area if deer are removed from that area.

10. Lyme disease is treated intravenously by use of tetracycline.

11. Epidemic and endemic are two phases of relapsing fever.

12. The causative microorganism of anthrax is a species of Pseudomonas.

13. There are respiratory and gastrointestinal forms of anthrax.

14. *Bartonella henselae* is probably a normal oral biota of cats and dogs.

15. *Rickettsia ricketsii* normally becomes concentrated in the alveoli of the lungs when it infects the body.

16. Early detection of RMSF is relatively easy.

17. Early diagnosis of typhus is difficult.

18. Only humans and lice are necessary to maintain typhus.

19. Murine typhus is endemic to Western Europe.

20. Typhus is caused by a virus.

B. The following statements are about viral infections. Label each of the following statements as true or false. If false, correct it.

1. Yellow fever is caused by *A. aegypti*.

2. A vaccine exists to prevent dengue fever.

3. Infectious mononucleosis occurs frequently in young children.

4. EBV typically infects the neurons of the body.

5. EBV establishes a latent infection in T lymphocytes.

6. The EBV can be oncogenic.

7. HIV infection usually begins in red blood cells.

8. From viral load a person has AIDS when the CD+4 T cell count drops to 600.

9. HAART contains reverse transcriptase and protease inhibitors.

10. HIV infection can increase the chance of developing certain kinds of cancer.

11. AIDS can be transmitted by the bite of an insect vector.

12. Only a small fraction of HIV infections are acquired in the hospital.

13. Great progress has been made in AIDS research since 1981.

14. The sole reservoir of the Ebola virus is the human population.

15. The Ebola virus belongs to the Filoviridae.

C. The following statements are about protozoan infections. Label each of the following statements as true or false. If false, correct it.

1. Most people who suffer from malaria are children.

2. After introduction by a mosquito bite, the *Plasmodium* is converted from a merozoite to a sporozoite in the liver.

3. The gametocytes of Plasmodium develop from merozoites in the human blood.

4. The release of malarial antibodies causes the periodic chills in the human.

5. The protozoan causing malaria is introduced into the human body through the bite of the *Culex* mosquito.

6. Today malaria is confined almost exclusively to tropical and subtropical areas.

7. Natural immunity to malaria is strong.

8. Tachyzoites multiply rapidly in mature red blood cells.

9. No vaccine exists for toxoplasmosis.

10. A tick vector transmits the parasite for babesiosis from one host to another.

D. The following statements are about helminth infections. Label each of the following statements as true or false. If false, correct it.

1. Shistosome eggs hatch directly into cercariae larvae.

2. Cercariae larvae directly infect people to cause shitosomiasis.

3. Shistosomes are prevalent in Asia, Africa, and the Middle East.

4. Filariasis is introduced into the human body through the bite of a tick.

5. In the development of elephantiasis, the circulation of the systemic blood flow is blocked.

Multiple Choice: Review

1. Pulmonary veins

    A. return blood to the heart from regions throughout the body.
    B. return blood to the heart from the lungs.
    C. transport blood from the heart to various body regions.
    D. transport blood from the heart to the lungs.

2. Endocarditis in humans is mainly

    A. bacterial and affects the inner layer of the heart.
    B. bacterial and affects the outer layer of the heart.
    C. viral and affects the inner layer of the heart.
    D. viral and affects the outer layer of the heart.

3. Trypanosomiasis is caused in humans by a

    A. bacterium.
    B. fungus.
    C. protozoan.
    D. virus.

4. *Yersinia pestis* is a

    A. Gram-negative coccibacillus.
    B. Gram-negative spirochete.
    C. Gram-positive coccibacillus.
    D. Gram-positive spirochete.

5. Occuloglandular disease infects mainly the

    A. brain.
    B. eye.
    C. heart.
    D. kidney.

6. Select the incorrect statement about *B. burgdoferi*.

    A. It causes Lyme disease.
    B. It is often recovered in blood of patients infected with it.
    C. It is a spirochete.
    D. It is easy to cultivate in the laboratory.

7. Anthrax is caused by a species of

    A. *Aerobacter*.
    B. *Bacillus*.
    C. *Pseudomonas*.
    D. *Staphylococcus*.

8. Select the incorrect statement about *Rickettsia rickettsii*.

   A. It causes RMSF.
   B. It is Gram-negative.
   C. It is rod-shaped.
   D. It mainly infects neurons.

9. Yellow fever is caused by a

   A. bacterium.
   B. fungus.
   C. protozoan.
   D. virus.

10. The nasopharyngeal carcinoma caused by EBV is common in

   A. Africa.
   B. China.
   C. India.
   D. South America.

Answers

Correlation Questions

1. From an Hh x Hh mating, the possible genetic recombinants among the offspring are: HH, Hh, Hh, and hh. Produced by a dominant gene, three-fourths of these recombinants produce an offspring with a genetic tendency to produce high cholesterol.

2. The left ventricle has a thicker layer of muscle, pumping blood through a longer systemic trip through the body compared to the pulmonary circuit. For this circuit, blood is pumped from the thinner right ventricle.

3. Capillaries are exchange vessels. Plasma from these blood vessels become tissue fluid. As this intercellular fluid has nutrients and oxygen to serve the nearby body cells (e.g. brain cells), a thin wall facilitates easy passage from the blood to the tissues. A capillary has one thin layer of epithelial cells, simple squamous epithelium.

4. When a given body region is infected, the circulating blood is a vehicle to spread it throughout the body. This makes the origin of the disease difficult to pinpoint. Its widespread distribution can also make the systemic infection difficult to treat.

The Cardiovascular and Lymphatic Systems

A. 1. left ventricle
   2. backflow
   3. more oxygen, less carbon dioxide

4. vein
5. capillaries
6. pericardial

Infections of the Heart

A. 1. endocardium
2. vegetations
3. *Staphylococcus aureus*
4. bacterial
5. Chagas'
6. liver, myocardial
7. enteroviruses
8. pericardectomy

Systemic Infections

A. 1. False; The reservoir is different species of rodents.
2. True
3. False; Pneumonic plague is more virulent.
4. True
5. False; They are phagocytized but not killed.
6. True
7. True
8. False; Arthritis develops six months after the tick bite.
9. True
10. False; Drugs used are doxycycline, amoxicillin, and erythromycin.
11. True
12. False; The causative microorganism is *B. anthracis*.
13. True
14. True
15. False; It is found along the inner lining of blood vessels.
16. False; Early detection is difficult.
17. True
18. True
19. False; It is endemic in the U.S.
20. False; It is caused by a bacterium.

B. 1. True
    2. False; A vaccine is not available.
    3. False; Infection usually does not occur until the age span of 15 to 25.
    4. False; If infects epithelial cells near the salivary glands.
    5. False; It produces the latent infection in B lymphocytes.
    6. True
    7. False; Surface (epithelial) cells, called dendritic cells, become infected first.
    8. False; It occurs when this count drops to 200.
    9. True
    10. True
    11. False; It is not one of the three modes of transmission.
    12. True
    13. True
    14. False; It is an unknown virus in the tropical rain forest.
    15. True

C. 1. True
    2. False; A sporozoite is converted to a merozoite.
    3. True
    4. False; The release of cytokines such as tumor necrotizing factor causes the chills.
    5. False; It is acquired from the bite of the *Anopheles* mosquito.
    6. True
    7. False; It is a weak response.
    8. False; They invade all body cells except mature red blood cells.
    9. True
    10. True

D. 1. False; They hatch into miracidia larvae.
    2. True
    3. True
    4. False; It is introduced by the bite of a mosquito.
    5. False; It blocks lymphatic circulation.

Multiple Choice: Review

1. B 2. A 3. C 4. A 5. B 6. B 7. B 8. D 9. D 10. B

Chapter 28
Microorganisms and the Environment

Key Terms

biogeochemical transformations
soil fertility
mycorrhizae
rhizosphere effect
eutrophic
biogeochemical cycle
root hair
nitrification
desulfurylation
trickling filter
methanogen
leach field
coliform bacterium

mineralization
nitrogen fixation
rhizosphere
R:S ratio
blooms
nitrogenase
ammonification
denitrification
biochemical oxygen demand
activated sludge
septic tank
sewage farming
MPN test

Study Tips

1. Answer the different kinds of questions at the end of the chapter. Answer them in your own words after studying the chapter. After answering them, then check the chapter content and compare it with your answers.

2. Review the chapters on basic chemistry (Chapter 2) and prokaryotic species (Chapter 11). How does the information in these chapters help you understand the concepts in Chapter 28?

3. Study the various biogeochemical cycles in this chapter. After studying them, try to outline each cycle in your own words and by your own patterns.

4. Is there a local plant that treats waste water? Visit it and write a short report on your findings. Talk to your instructor about making this an extra credit project.

Correlation Questions

1. Farmers practice crop rotation—planting nitrogen-fixing leguminous plants during certain years. Why is this a good strategy?

2. Why does the surface layer of the soil often have a dark, moist makeup?

3. What gas buildup in the atmosphere can cause the greenhouse effect. How does it produce this effect?

4. Where is the place for the biosphere in the levels-of-organization hierarchy?

Life and the Evolution of Our Environment

A. Label each of the following statements as true or false. If false, correct it.
1. Sulfur is a usable, major biological element.
2. The Earth was formed about 3.5 billion years ago.
3. The early atmosphere of the Earth lacked oxygen.
4. The oldest fossils known resemble protozoa.
5. Today oxygen comprises about 30 percent of the atmosphere of the Earth.

Microorganisms in the Biosphere

A. Complete each of the following with the correct term.
1. The _____ is the region of the Earth that supports life.
2. The process of _____ converts organic material to inorganic material.
3. Soil fertility depends on an adequate supply of the elements _____, _____, and _____.
4. Most soils contain _____ microorganisms, as the soil can become hot during the day.
5. Actinomycetes are Gram-_____ bacteria.
6. The bacterium _____ _____ is an opportunist that causes infections in burn victims.
7. _____ is a symbiotic relationship that benefits both organisms involved.
8. _____ is a symbiotic association between fungi and the roots of plants.
9. The _____ ratio is the ratio of the concentration of microorganisms in the rhizosphere to their concentration in the adjacent soil.

10. *Pseudomonas* has Gram-_____ species.
11. About __(number)__ percent of the water of the Earth is seawater.
12. Most photosynthesis is conducted by _____ in the water.
13. _____ waters are enriched with nutrients.
14. Compared to the soil and water, croorganisms do not grow in the _____.
15. _____ are tiny particles of liquid.

The Cycles of Matter

A. Match each of the following descriptions to the correct biogeochemical cycle.
   1. ammonia is converted
   2. its gas is a small part of the air
   3. gaseous intermediate is lacking
   4. involves tenth most abundant element
      A. carbon cycle
      B. nitrogen cycle
      C. phosphorus
      D. sulfur

B. Complete each of the following statements with the correct term.
   1. Nitrogen makes up about _____ percent of the atmosphere of the Earth.
   2. The process of _____ _____ converts nitrogen from the atmosphere into a usable form.
   3. _____ is an enzyme catalyzing nitrogen fixation.
   4. The _____ process industrially fixes nitrogen.
   5. Nitrification converts ammonia to the _____ ion.
   6. _____ is the main form of nitrogen used by plants.
   7. _____ is the conversion of nitrate into nitrogen gas.
   8. _____ destroys the ozone layer of the atmosphere.
   9. Carbon dioxide makes up about __(number)__ percent of the atmosphere of the Earth.
   10. Carbon is present in the atmosphere mainly as _____.
   11. _____ are organisms that generally fix carbon dioxide from the atmosphere by photosynthesis.
   12. _____ is the metabolic process that returns carbon dioxide to the atmosphere.
   13. Carbon is dissolved in waterways as the _____ion.
   14. Phosphorous is present in the ocean as the _____ion.
   15. Phosphorus lacks a _____ intermediate as it cycles.
   16. In short supply, phosphorus is a _____ nutrient in most soils.
   17. The main loop of the sulfur cycle consists of reducing sulfate ions into the compound _____.
   18. Most microorganisms derive their sulfur from the _____ ion.
   19. By _____ organic sulfur compounds are converted to hydrogen sulfide.
   20. Sulfur-oxidizing bacteria oxidize sulfide to _____.

Treatment of Waste Water

A. Label each of the following statements as true or false. If false, correct it.
   1. Sewage is municipal waste water.
   2. Sewage treatment plants add oxygen to sewage at a slower rate than microbes remove it by metabolism.
   3. Primary sewage treatment is biological rather than mechanical.
   4. Secondary sewage treatment is mechanical rather than biological.
   5. The effluent is the liquid entering the sewage treatment plant.
   6. Digested sludge is the solid portion of primary sludge remaining after primary treatment.
   7. Most tertiary treatments of waste are chemical rather than biological.
   8. A septic tank is a small anaerobic digester.
   9. The final effluent from a properly functioning sewage plant has a low BOD.
   10. Methanogens use methane.

Treatment of Drinking Water

A. Complete each of the following statements with the correct term.
   1. To clarify water is to remove _____ _____ from it.
   2. Ozone or _____ is added to water supplies to kill microorganisms.
   3. _____ bacteria are used as indicators for the quality of drinking water.
   4. The first test of the MPN test is the _____ test.
   5. By the MF procedure microbes are cultured on a(n) _____ plate.

Multiple Choice: Review

1. Select the correct description for humus.
   A. inorganic, brown or black
   B. inorganic, red or yellow
   C. organic, brown or black
   D. organic, red or yellow

2. Select the correct description of most fungi.
   A. aerobic, decomposers
   B. aerobic, producers
   C. anaerobic, decomposers
   D. anaerobic, producers

3. Which kind of symbiotic relationship benefits both organisms?

   A. commensalism
   B. mutualism
   C. opportunism
   D. parasitism

4. Water covers about _____ percent of the surface of the Earth.

   A. 50
   B. 60
   C. 70
   D. 80

5. The conversion of nitrates into nitrogen gas is

   A. ammonification.
   B. denitrification.
   C. nitrification.
   D. nitrogen fixation.

6. Carbon dioxide is returned to the atmosphere by

   A. photosynthesis.
   B. respiration.

7. The main inorganic supply of phosphorus is in

   A. rocks.
   B. the atmosphere.
   C. the ocean.
   D. the streams.

8. Desulfurylation converts organic sulfur compounds to

   A. hydrogen sulfide.
   B. sulfate.
   C. sulfuric acid.
   D. sulfuric acid.

9. The primary treatment of waste water is usually

   A. biological.
   B. chemical.
   C. enzymatic.
   D. mechanical.

10. The second step of the MPN procedure is the _____ test.

   A. coliform
   B. completed
   C. confirmed
   D. presumptive

Answers

Correlation Questions

1. Some species of crops deplete nitrogen from the soil, fulfilling a need in their metabolism. Through nitrogen fixation, leguminous plants replace this element that is depleted in the soil.

2. The dark, moist layer is humus. Humus is decaying organic matter.

3. The greenhouse effect results from the buildup of a carbon dioxide in the air. It traps heat onto the surface of the Earth, warming it.

4. The biosphere is a thin layer surrounding the Earth supporting life. It is the largest level among levels of organization.

Life and the Evolution of Our Environment

A. 1. True
    2. False; The Earth was formed about 4.6 billion years ago.
    3. True
    4. False: The oldest known fossils resemble cyanobacteria.
    5. False; Today oxygen comprises about 21 percent of the atmosphere of the Earth.

Microorganisms in the Biosphere

A. 1. biosphere 2. mineralization 3. sodium, potassium, phosphorus 4. thermophilic 5. positive 6. *Pseudomonas aeroginosa* 7. mutualism 8. mycorrhizae 9. R:S 10. negative 11. ninety-nine 12. phytoplankton 13. eutrophic 14. air 15. aerosols

The Cycles of Matter

A. 1. B 2. A 3. C 4. D

B. 1. eighty 2. nitrogen fixation 3. nitrogenase 4. Haber 5. nitrate 6. nitrate 7. denitrification 8. denitrification 9. 0.03 10. carbon dioxide 11. autotrophs 12. respiration 13. carbonate 14. phosphate 15. gaseous 16. limiting 17. hydrogen sulfide 18. sulfate 19. desulfurylation 20. sulfate

Treatment of Waste Water

A. 1. True
    2. False; They do the opposite of this.

3. False; It is mechanical rather than biological.

4. False; It is biological rather than mechanical.

5. False; It is liquid leaving the treatment plant.

6. True

7. True

8. True

9. True

10. False; They are bacteria that make methane.

Treatment of Drinking Water

A. 1. pathogenic microorganisms 2. chlorine 3. coliform 4. presumptive 5. EMB

Multiple Choice: Review

1. C 2. A 3. B 4. C 5. B 6. B 7. A 8. A 9. D 10. C

Chapter 29
Microbial Biotechnology

Traditional Uses of Microorganisms
    Lactic Acid Bacteria
    Yeasts
    Mixed Cultures
Microbes as Insecticides
Microbes as Chemical Factories
    Anaerobic Fermentation
    Aerobic Processes
    Chemical Processes
Using Genetically Engineered Microorganisms
    Medical Uses
    Agricultural Uses
Summary

Key Terms

industrial microbiology
silage
ripening
enology
brandy
blackstrap molasses
impeller
intron
transformation

biotechnology
curdling
whey
viticulture
saccharified
sparger
penicillinase
reverse transcriptase

Study Tips

1. This chapter shows the application of many concepts that you have learned throughout your study of microbiology. Can you identify these concepts? Review the text on topics such as bacterial metabolism and bacterial genetics.

2. Visit a local brewery or bakery to learn more about the application of microbiology in commercial industry.

3. Does your college have a biologist conducting research that is relevant to genetic engineering? Interview this person to learn more about this topic firsthand.

4. Start an Internet search with the term *biotechnology*.

Correlation Questions

1. Is the refrigeration of milk a sterilization procedure or is there another strategy for its use?

2. How has biotechnology helped treat people with pituitary dwarfism?

3. How has biotechnology helped treat people with diabetes mellitus?

4. A protease inhibitor is added to a mixture of rennin and protein. How will this affect the chemical activity of the mixture?

Traditional Uses of Microorganisms

A. Complete each of the following statements with the correct term or terms.
   1. The production of acid by lactic acid bacteria _____ food, as their acid kills or inhibits other microorganisms.
   2. The term "sauerkraut" means _____.
   3. Lactic acid bacteria of *Lactobacillus* _____ ferment sugar in plant material.
   4. Cheese is made by the steps of _____ and _____.
   5. Cottage cheese is made from the action of the enzyme _____.
   6. Hard cheese is softened by the chemical digestion or _____ of the proteins in the cheese.
   7. Yeasts of the scientific name _____ _____ are used to ferment sugar.
   8. _____ is the process of winemaking.
   9. _____ is the process of grape growing.
   10. Either _____ or _____ wine can be made from red grapes.
   11. The metabolic products of the bacterium _____ _____ give sherry its distinct taste.
   12. Botrytis cinerea is a _____ that produces certain sweet wines.
   13. By secondary fermentation, the bacterium *Oenococcus oeni* converts _____ acid to lactic acid for winemaking.
   14. The enzyme amylase produces a disaccharide from starch. This disaccharide, consisting of bonded glucose molecules, is _____.
   15. Beer is flavored from the flowers of _____ _____.
   16. Vinegar is a solution of _____ _____.
   17. Vinegar is produced by the action of the bacterium _____.
   18. _____ _____ is the gas produced by the yeast *S. cerevisiae* to leaven bread.

Microbes as Insecticides

A. Complete each of the following statements with the correct term.
   1. Bt is produced by the bacterium _____ _____.
   2. Bp is produced by the bacterium _____ _____.
   3. Spores of the protozoan _____ _____ are sold commercially as a bait to combat grasshoppers.

Microbes as Chemical Factories

A. Label each of the following statements as true or false. If false, correct it.
1. Fermentations produce large amounts of ATP.
2. The most common substrates for industrial fermentations are blackstrap molasses.
3. Methanol is the alcohol found in alcoholic beverages.
4. Gasohol is 50 percent gasoline and 50 percent alcohol.
5. A species of the genus *Clostridium* has been used commercially to make the solvents acetone and butanol.
6. Most modern industrial microbiological processes are anaerobic.
7. The antibiotic industry began with the work of Alexander Fleming.
8. Penicillin is difficult to synthesize chemically.
9. Most broad-spectrum antibiotics are synthesized commonly in industry.
10. Lysine is added to bread in the United States.
11. Corticosteroids produce an inflammatory response in the human body.
12. Microorganisms are a rich source of enzymes that have many commercial uses.

Using Genetically Engineered Microbes

A. Label each of the following statements as true or false. If false, correct it.
1. hCG is secreted from the thyroid gland.
2. Prokaryotes lack enzymes to eliminate introns.
3. cDNA is produced from reverse transcription.
4. As a technique of genetic engineering, transformation is the introduction of native or recombinant DNA into a host cell.
5. The action of the hormone insulin is to lower glucose in the cell.
6. Human insulin can now be produced from *E. coli*.
7. tPA binds to blood clots and dissolves them.
8. tPA is produced from microbial cells by genetic engineering.
9. Human DNAase is used to treat cystic fibrosis.
10. Streptokinase is used to treat victims of heart attacks.
11. *P. fluorescens* is a species of bacterium that inhabits the rhizosphere.
12. The presence of *P. syringae* makes frost damage less likely on plants.

Multiple Choice: Review

1. Select the correct statement about lactic acid bacteria.
   A. They are harmful to humans and decompose food.

B. They are harmful to humans and preserve food.

C. They are harmless to humans and decompose food.

D. They are harmless to humans and preserve food.

2. Hard cheese is made softer by the

A. formation of proteins through dehydration synthesis.

B. formation of proteins through hydrolysis.

C. digestion of proteins through dehydration synthesis.

D. digestion of proteins through hydrolysis.

3. *Saccharomyces cerevisiae* is a

A. bacterium that ferments sugars.

B. bacterium that makes sugars.

C. yeast that ferments sugars.

D. yeast that makes sugars.

4. Select the incorrect association.

A. *Acetobacter*/virus

B. amylase/hydrolyzes starch

C. saccharified/produces glucose

D. vinegar/acetic acid

5. Species of the bacterial genus _____ are insect pathogens.

A. *Aerobacter*

B. *Bacillus*

C. *Escherichia*

D. *Pseudomonas*

6. Fermentations produce _____ amounts of ATP.

A. large

B. small

7. Gasohol is ___1__ percent gasoline and __2__ percent alcohol.

A. 1 - ninety, 2 - ten

B. 1 - seventy, 2 - thirty

C. 1 - fifty, 2 - fifty

D. 1 - twenty, 2 - eighty

8. In anaerobic processes in industry, a rotating shaft is a(n)

A. fermenter.

B. impeller.

C. silager.

D. sparger.

9. Penicillase is

A. alpha-lactamase that destroys penicillin.

   B. alpha-lactamase that makes penicillin.
   C. beta-lactamase that destroys penicillin.
   D. beta-lactamase that makes penicillin.

10. A deficiency of insulin in the body will tend to _____ the concentration in the blood.
   A. decrease
   B. increase

Answers

Correlation Questions

1. Refrigeration provides a temperature that is too low for many bacteria to carry out a significant rate of metabolism. This includes the bacteria that produce lactic acid that sours milk. However, it does not remove them from the milk by sterilization.

2. The growth hormone can be made is larger amounts. It is used is pituitary dwarfs, replacing the hormone that they cannot produce naturally.

3. Insulin can be made in larger amounts. More is available to treat more diabetics.

4. The inhibitor stops the protease activity of rennin, curtailing its activity.

Traditional Uses of Microorganisms

A. 1. preserves
    2. acid cabbage
    3. *plantarum*
    4. curdling, ripening
    5. rennin
    6. hydrolysis
    7. *Saccharomyces cerevisiae*
    8. enology
    9. viticulture
  10. red, white
  11. *S. fermentati*
  12. fungus
  13. malic
  14. maltose
  15. *Humulus lupulus*
  16. acetic acid

17. *Acetobacter*

18. carbon dioxide

Microbes as Insecticides

A. 1. *B. thuringiensis*

2. *B. popilliae*

3. *Nosema locustae*

Microbes as Chemical Factories

A. 1. False; Fermentations produce small amounts of ATP compared to aerobic respiration.

2. True

3. False; Ethanol is the alcohol of alcoholic beverages.

4. False; Gasohol is 90% gasoline and 10% alcohol.

5. True

6. False; Most of these processes are aerobic.

7. True

8. True

9. False; They are too complex for chemical synthesis.

10. False; It is not added, as enough is available from the consumption of other foods.

11. False; They suppress inflammation.

12. True

Using Genetically Engineered Microbes

A. 1. False; It is secreted from the anterior pituitary gland.

2. True

3. True

4. True.

5. False; It lowers the glucose concentration in the blood.

6. True

7. True

8. False; It is produced from CHO cells.

9. True

10. True

11. True

12. False; Its presence makes frost damage more likely.